THE COMPLETE CNC MASTER HANDBOOK

A Comprehensive Guide to Operation, Setup and Programming Your Blueprint to Success

By

Nishioka Yoshihiro

TABLE OF CONTENTS

- DEMO 3
- INDEX 10
- HOW TO START CNC MACHINE 14
- WHAT IS CNC MACHINE 17
- CONTROL PANEL (PART 1) 20
- CONTROL PANEL (PART 2) 30
- OFFSET PROCEDURE 41
- CAREER GUIDANCE 1 49
- JOB SETTING 53
- TOOL SELECTION AND ITS NOMENCLATURE 67
- MEASURING INSTRUMENTS 85
- CAREER GUIDANCE 2 92
- G-CODES 94
- M-CODES 104
- STANDARD FORMAT 107
- NOSE RADIUS COMPENSATION 114
- FINDING CO-ORDINATES 120
- EXAMPLES (PART 1) 128
- EXAMPLES (PART 2) 144
- CYCLES (PART 1) 167
- CYCLES (PART 2) 197
- MORE INFORMATION 232
- CAREER GUIDANCE 3 241

DEMO

we have covered CNC operating CNC setting and CNC programing in detail and not only theoretical knowledge, but also with the help of simulator we have given practical knowledge as well. So see this video so that you can understand how we have teach you in this course. So our third cycle is drill cycle. And for this cycle we use G70 for code. So in this cycle also you will get two blocks. So in our first block we have G70 four and R. So these are each retraction amount. So if you know how this drilling works it do not take hold good in one time. Firstly two will take light cut. Then it move back. Then again it take good and again move back. So in this way tool makes that drill. So that backward distance is known as retraction amount. That means tool will take good and then move back. So that backward distance we write here and in second block we write G70 for x. So this x is diameter up to which drilling has to be done means up to which diameter we are going to take. Good. But we don't do. Drilling in diameter means we don't do drilling in X direction. We do drilling in Z direction. So that's why we don't write this x in drill cycle. After that we have z means distance up to which drilling has to be done, means up to how much distance we are going to make that drill.

Drill Cycle(G74)

(First Block)
R----Retraction amount

G74 R_

(Second Block)

G74 X_ Z_ P_ Q_ F_

X----Diameter upto which drilling has to be done
Z----Distance upto which drilling has to be done
P----Depth of cut in X axis

(0.1*1000 = 100)

Q----Depth of cut in Z-axis (in microns)
F----Feed

(We have used 40mm drill)

So that Z distance we have to write here, be it depth of cut in x axis. But as we are not going to take good in x axis, that's why we do not write this p in drilling cycle as we do grooving in x axis and drilling in z axis. That's why we don't write this p here. After that q depth of cut in z axis. So as we do drilling in z axis. So here we have to write depth of cut and that depth of cut. We have to write in microns. So I have already told you this m m to micron convergence. So here you have to convert depth of good in microns. And that value you have to write here. So suppose we are going to take zero point 1MM depth of cut. And to convert m into microns we have to multiply it by 1000. So if we multiply 0.1 m by 1000 then we get 100 microns. So you have to write 100 here. And after that we have feed. So this feed you already know. So here we have used for the m m drill. Now we will see our drill cycle. So here I have written one simple program in which I have included G70 for cycle. So as you can see G70 for ARM. So I have taken our value add 1.0 means tool will retract by 1MM in next block. We have G70 for here. X will not come as we don't do drilling in X, we take good in Z and we will take good in Z up to -80.0. And that's why I have written here z -80.0. After that p so p it depth of cut in x as we are not going to take good in x. So this p also not come here q is depth of cut in z. So here I have taken two m cut. So we convert to M into microns. So that will come 2000. And after that feed 0.1. So now I will run this program. So see whether our G70 for cycle makes drill or not. So see, tool is not cutting material in one cut only.

It is moving back and forth to cut that material means our G70 for cycle is working here. So this is all about value control. So as you see in video, you are going to learn theoretically as well as practically. So in this course we have covered two control panels. So video that you have recently seen in that we have done programing with Fanuc panel. Now see video with Siemens control. And now we will make simple group here. So for that we will keep first position and second option is like this. If you want this type of group then select second option. So you have to choose this according to your requirement. But we want simple group. So select first option. And in that you have to enter here values. So see if you already know this then we will make roughing as well as finishing both with this same tool. So x zero and z zero means we have to give dimensions at this point. So here we will start doing group from last means. Firstly we will make last group, then middle and then first. In this way we will make this group. Normally we make group from starting. But here we will do something new. We will make these group from last.

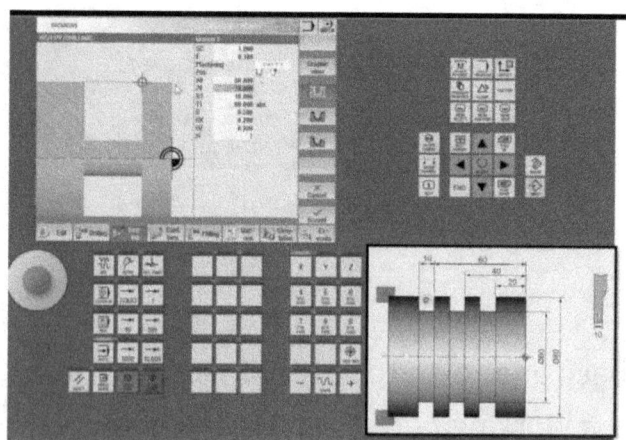

So that round part will come here. And we have here x 80. And if we make it Z then it will be 60 means -60.0. So we will put these values in x zero and z zero. After that we have group width b one means group width. So group width 810 m. So right then here after that t one means group depth diameter. So here we want this group up to diameter 60. So we will write this 60 in T one. After that we have depth of cut. So we will take 0.5 depth of cut. After that allowances in x direction. Allowances in z direction. So write these values and n means number of groups means how much grooves you want to make. So here we are going to make three group. So we will write n is equal to three. After that it is asking DP means distance between this point to this point. So how much distance it is. So if you can see here. So from this point to this point distance here is 20 m. So if you calculate you will get this distance as 20. And after that press exit. So here we have given all the information. After that we will lift our tool up to x 100. After that we will do two homing and then M30. End of program. So in this way you have to write this groove cycle in Siemens control. So let us see whether it really works or not. So see firstly it is making last groove. After that middle one, as we are making this groove from last to first. And then first groove. So here our all three grooves are ready. That means our cycle is accurate. So in this course you are going to learn programing with FANUC control panel as well as Siemens control panel. We have seen many people who work 8 to 10 years on CNC, and then also they are not able to become CNC programmer as they don't know how to move ahead in career. And that is why in this course, we have added three career guidance videos so that you will know the growth path of your career means from CNC operator, how to become CNC, etc. and from CNC center how to become CNC programmer. And not only this, we have also told you how to increase your salary means how you can become a good CNC programmer within one year, and how to increase your salary after becoming a programmer. So that everything we have covered in these three career guidance videos means this is complete course. Firstly, we will make you a good CNC programmer and also we will give you knowledge about how to increase your salary. You will become a good CNC programmer only then when you know every small thing in detail as you are, one small mistake will reject your whole job. And that is why people are not able to become a programmer. Even after working for 8 to 10 years on CNC. And you won't get this complete knowledge from any videos or from YouTube videos. For this complete knowledge, you require a proper training and then only you will become a good programmer. YouTube videos or other videos will definitely give you extra knowledge. But for complete knowledge you require a proper training or proper course. We have done many surveys and from that we get to know that many people want to learn programing, but either they do not have time to join offline classes and also offline institutes charges much higher fees and to pay such high amount of face is not possible for everyone. So by considering all these things, we have created this course. This course is online, so you can watch these videos at any time and learn from it. And we thought a lot about fees and we tried to keep as low price as we can so that everyone can learn programing with the help of this course. So if any helper also works for ten days, then also he can able to buy our course very easily. So if you are getting all this knowledge instead of your ten days of work, then you must take it as much you invest on yourself, on your knowledge. The faster you will grow in your career. So you must take it and move ahead in your career.

INDEX

This will yield an exceptional outcome for you. Initially, we will examine the index, which denotes the content that we will be studying in this course. Therefore, the initial subject of this course is the process of initiating a C and C machine. Therefore, it is evident that we will commence our education with the fundamentals.

SR.NO	TOPIC
1	HOW TO START CNC MACHINE
2	WHAT IS CNC MACHINE
3	CONTROL PANEL
4	OFFSET PROCEDURE
5	SETTING
6	TOOL SELECTION AND NOMENCLATURE
7	MEASURING INSTRUMENTS
8	G-CODES
9	M-CODES
10	STANDARD FORMAT
11	NOSE RADIUS COMPENSATION
12	FINDING CO-ORDINATES
13	EXAMPLES
14	CYCLES
15	MORE INFORMATION

Therefore, even if you are unfamiliar with CNC and are new to this profession, you will be able to program with this course. Consequently, we will acquire knowledge at every level, from the fundamentals to the most advanced. Initially, we will examine the process of initiating a C and C machine. Afterward, we will examine the definition of a

CNC machine. Therefore, what is the precise nature of the CNC machine that we will be studying in this article? Control Panel will be the subject of our third topic. Therefore, this topic will provide a comprehensive examination of the Control Panel. We will then observe the agitated procedure. This implies that we will examine the process of causing anger in this section. Additionally, we will acquire the ability to determine the inverse value. In conclusion, this is the entirety of the subject matter that we will examine. The subsequent step will involve the establishment of a topic. That is to say, the first four topics are crucial for the operational means. Instruments for measuring the first four topics and this seventh topic. Therefore, these five subjects are specifically designed for the purpose of CNC operation. The initial four topics and this seventh topic. The following step will involve the acquisition of knowledge regarding the means of setting. This subject is extremely beneficial for the development of a C and C setup. Also, this fifth and sixth topic are specifically designed for the C and C set of means, in addition to this instrument, setting, and nomenclature. The first four and seven topics pertain to operating, while the fifth and sixth topics pertain to C and C settings. And our programming will commence with the eighth topic. Therefore, the eighth topic will feature G codes, while the ninth topic will feature M codes. Subsequently, we will examine the standard format for programming in the tenth topic. We adhere to a single standard format.

Additionally, it is necessary to compose the program using this standard format. So, in this topic, we will be learning the standard format. Compensation for nasal radius will be our subsequent subject. Therefore, this is the programming component. So, we will also observe this programming component subsequent to the coordinate determination. Therefore, this is a straightforward subject. However, if any pupil is unable to determine coordinates. Therefore, we have included this topic for that student, and you will find it to be of great assistance in the development of your programs. Following this, we will examine a few examples in this example topic. All operations, including drilling, threading, and grooving, will be observed. Additionally, each operation will be illustrated with a single example. Additionally, we will develop a program for that example. So this topic is of great significance for programming purposes. Once you have acquired all of the knowledge from the first to the 12th topic, this topic will be straightforward for you, as we implement all of the knowledge from these 12 topics in this chapter. Therefore, we will observe numerous examples of various categories and develop a program to accommodate them. Therefore, in this subject, we will employ G00 and G01 protocols to develop a program. Following this, we will observe cycles. Consequently, we will develop programs for a variety of examples using cycles. Therefore, there are distinct cycles in CNC programming. Therefore, this topic will cover all of the cycles. Therefore, our programming syllabus will conclude with this subject. Following that, we will have additional information. We will provide you with an external website In this career guidance project, we will provide you with some guidelines on how to apply your operational knowledge to a machine and in which direction to advance in your career afterwards. Therefore, we will furnish you with this comprehensive guidance in our inaugural career guidance project. It is imperative that you possess a comprehensive understanding of these concepts in order to develop into a proficient CNC setter, CNC programmer, and CMC operator. Therefore, you will be able to quickly and effectively acquire the skills of operating with the assistance of this career guidance project. This is also true for the location. Therefore, you will receive this career guidance project after the sixth topic. In this project, we will offer comprehensive guidance on how to become a setter in a shorter amount of time. Additionally, which aspect necessitates additional attention? So, we will provide you with all of this guidance in this project, and subsequently, we will provide you with a third piece of career advice. The project in the final section signifies that we will furnish you with career guidance project following the conclusion of our course. In the career guidance project, we will provide all the necessary information to advance in your career. In this final career guidance project, we will provide you with information on how to implement your programming knowledge, increase your salary, and more. Therefore, we will offer you these three career guidance projects in this manner. Therefore, you will acquire knowledge and the ability to advance in your career. That knowledge will also be acquired, and it will enable you to accomplish your professional objectives in a significantly shorter amount of time.

HOW TO START CNC MACHINE

A Comprehensive Guide I am aware that this subject is straightforward; however, there may be some of us who are new to this discipline. Therefore, it is imperative that we bring all of these individuals with us. Therefore, what is the process for initiating a CNC machine? This may not be necessary if you are employed by a large-scale organization. There are fewer possibilities of you turning off the machine, as we have three operational vehicles. However, if you are employed in a small-scale company, you may require this information. This is because, in the event that you are the sole employee of the company, you must be aware of how to operate the CNC machine. Initially, it is necessary to activate the primary switch. This switch is the primary switch, as displayed here. Therefore, it is necessary to activate the switch. The electricity supply will commence immediately upon the activation of the primary switch. Therefore, it is necessary to proceed to your C and C machine after activating the primary switch. A receptacle similar to this one is located in close proximity to the CNC machine. Therefore, it is necessary to activate this device as well. You may now inquire as to the nature of this container. Therefore, as you are aware, electricity is not consistently reliable. Despite fluctuations in electricity, a stable electricity supply is necessary for CNC machines.

2.Switch On The Stabilizer

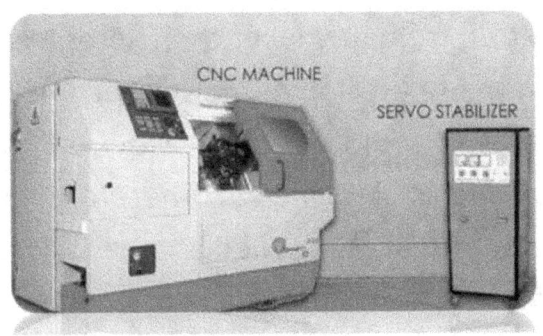

So in order to transform this fluctuating electricity into stable electricity, we employ this receptacle, which is referred to as a stabilizer. Therefore, the voltage supply is maintained in order to convert fluctuating electricity into stable electricity in this stabilizer. The disadvantages persist. Therefore, this stabilizer is also referred to as a voltage stabilizer. Therefore, it is necessary to activate this stabilizer. Subsequently, click the NC icon. Therefore, this is the appearance of the CNC on the button. Therefore, this is the method by which you can initiate your CNC machine. The homing procedure must be conducted after the CNC machine is initiated. We will be studying this localization process in our forthcoming topics. For the time being, it is important to remember that homing is required after the CNC machine is initiated. Therefore, we have acquired the knowledge necessary to initiate a CNC machine. Immediately execute one action. If you are currently unemployed, commence your CNC operation from any location. Therefore, commence with the operation of a CNC machine; as you progress through this course, you will acquire both theoretical and practical knowledge. Therefore, in order to implement this practical knowledge on a CNC machine, a C and C machine is necessary. Therefore, commence your CNC operation as soon as feasible in order to implement all of this knowledge on your C and C machine. You will obtain a CNC operating position with ease. Start your employment search in a small-scale industry, as it is easier to learn in a small-scale setting than in a large-scale one. Therefore, commence your investigation.

WHAT IS CNC MACHINE

The subsequent subject is the definition of a CNC machine. Therefore, in CNC, C represents computer, N represents numeric L, and C represents control. Therefore, it is computer numerical control. This implies that we will utilize a computer to control the movements of all devices. In addition to the movements, there are various operations. Additionally, it is important to note that computers are incapable of comprehending languages such as English; they are limited to programming languages. Before the development of CNC machines, all of these operations, such as turning, drilling, and grooving, were performed manually with the assistance of the Lay the Machine. However, the Lay the Machine required a significant amount of time and effort to perform machining or these types of operations. As time progressed, the machine was equipped with new technologies, which is how our CNC machine was developed. Therefore, there are three configurations in this CNC machine. The initial step is the establishment of employment holding. Therefore, this is the configuration of our job holding system. Additionally, these are the positions.

MACHINE SETUP:

- Job Holding Setup:

This is the manner in which our labor is suspended between tasks. This will be thoroughly examined. However, for the time being, please examine the contents. The subsequent setup is a tool holding configuration in which all of the tools are installed. Additionally, we will utilize these instruments to reduce the material of the workpiece. Therefore, this is the comprehensive appearance of our workpiece. This configuration is referred to as a turret. Therefore, it is important to recall the name of the turret in which all of the instruments are installed. Our third configuration, which includes a control console. This is the most critical configuration for the execution of various operations and instrument movements. Therefore, there are numerous organizations that produce this control panel, including Simmons, Haas, and Fanu. In this course, we will be studying the Simmons control panel and the farm appearance, which are primarily employed in the industrial sector. This is analogous to our mobile phone, as there are numerous corporations that produce mobile phones, including Apple and Samsung. However, the inherent functionalities of their mobile phones are identical, despite the dissimilar loop structure. For example, the "calling" icon initiates a call, while the "message" icon initiates a message. It is precisely the same in terms of control panel structure; although the appearance may vary, the inherent functionalities are identical. Therefore, by mastering a single control panel, you will be able to operate any control panel with ease. We have now concluded our second topic. These two subjects are merely introductory. Therefore, we will commence our primary syllabus with the subsequent topic.

CONTROL PANEL (PART 1)

Control Panel is the third subject we will cover. Therefore, we will be able to execute a variety of operations, such as drilling, turning, and grooving, with the assistance of the Control Panel. In addition to these operations, we will be able to conduct tool movements. Therefore, this is our control interface. It is composed of three components. The initial component is the display, which provides a comprehensive overview of the tool's status, including its program, pace, distance, and other relevant information. The second component is the programming interface. Therefore, the programming interface enables us to compose, modify, and execute all program-related tasks. The operating interface is utilized in the third section. We are capable of performing two movements and all operational-related tasks. In this section, we will examine the Fanuc Control Panel. By acquiring a comprehensive understanding of the FANUC control panel, it is possible to operate any control panel with ease. Therefore, the reset key is the most frequently used key in the control panel, similar to the way it is used on our computer. In order to refresh our computer, we possess a refresh key. In order to refresh the CNC, we have implemented a reset command. However, this is not the sole application of the reset function. In the event of an emergency, the reset function is also employed to halt the machine. Emergency stops are also permissible in this location. The reset key is also employed to eliminate an alarm or message. It is inevitable that errors will occur when operating machinery, and no individual is entirely faultless. Therefore, the machine will inform you that you have made an error through the use of a message. Therefore, it is necessary to rectify the error and subsequently select the reset button. The message will be removed from the CSC immediately upon pressing the reset key. And if this message persists, it indicates that the issue or error has not been resolve. So, resolve the issue first and then select the reset key to eliminate the message. The third application is to resume the program, which involves initiating the program from the beginning. So, let us now determine whether it is functioning properly.

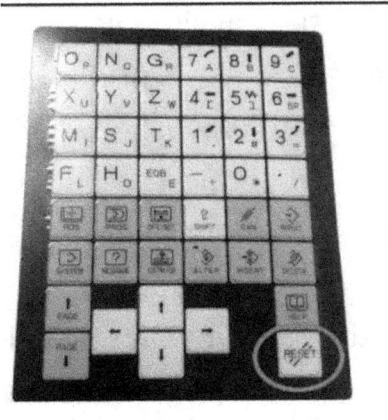

RESET

-To stop m/c in case of emergency.
-Remove alarm or message from screen
-To restart program

With the assistance of project. Therefore, it is evident that the spindle is in motion. This implies that our CNC machine consumes. Immediately upon pressing the reset button, our machine halts. Therefore, it is operational. In the event of an emergency, it is possible to utilize the reset function. Therefore, this pertains to the restore function. Page interface keys comprise our subsequent keys. Minsky's are employed to perform page-related tasks. Therefore, we will examine each item individually. Therefore, our initial character is the magician's key. This is the appearance of our position key. Now, let us examine the precise function of this position key. Therefore, it indicates the tool's position in relation to task zero. We will discover the nature of this position zero. However, for the time being, it is important to remember that the position key indicates the tool's position in relation to task zero. Additionally, it indicates the cutting tool's position. Therefore, this key is primarily employed to visualise the tool's position. We will now examine the position key in greater detail with the assistance of a project. Therefore, it is evident that our spindle is in motion. This is our operating interface. Therefore, upon pressing the position key, you and W will be obtained. Our tool's x and z distances from assignment zero are illustrated here. Now, closely observe this distance and retain it. I observe this distance change immediately upon commencing my cycling journey. This distance indicates that our instrument is capable of moving an additional 87 meters in the Z direction. However, as demonstrated here, this distance is exceedingly diminutive. This implies that our instrument will undoubtedly collide with the spindle. I have purposefully induced this situation in order to demonstrate the significance of this role. Key. I observe that the cycle tool collides immediately upon commencing operation, and the CNC machine generates an alarm. Therefore, this key will be of great assistance during the setup process, as it is necessary to determine the distance between the instrument and the task site and the actual remaining distance.

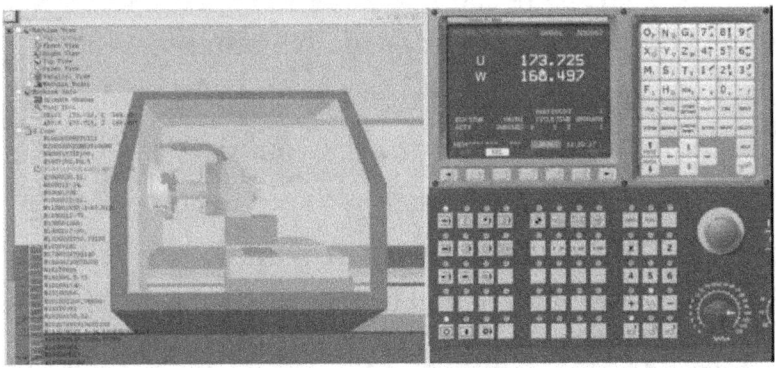

Consequently, this key will enable you to prevent an accident. Therefore, this is a critical component of the setting process. The subsequent key is the program key.

Therefore, this is the appearance of our program key. Therefore, it is employed to monitor the program. This implies that you are simultaneously executing programs and utilizing your stored programs. This implies that this key enables you to access any program stored in your CNC machine. All right. We will now examine this key in greater detail with the assistance of a project. Therefore, upon pressing the program key, the depicted program will be our current operating program. Therefore, the FANUC control program name contains 15 numbers. It commenced with the letter O. There are four numerals following O. This is a control rule that mandates that the program name begin with the letter "O." Subsequently, you are required to provide four numbers. However, Siemens is not responsible for this. You have the option to assign any name to your program in Siemens, beginning with an alphabet or number. Anything you are able to provide. All right. Therefore, in Fanuc, I can locate a single number program. Therefore, I am required to compose or 0001. And then I must strike this key. Our single-digit program is exclusively visible on the screen. It is important to keep in mind that our auto mode must be enabled in order to search for any program. These modes will be observed at some point. However, for the time being, it is important to remember that auto mode must be enabled in order to search for any program. Therefore, we will investigate the feasibility of executing this 50-number program. Therefore, activate auto mode and enter 0050 or write. Subsequently, utilize this key to initiate the application. And then, simply select the "cycle start" key. Observe the simplicity of our program in operation. Locate the program and initiate the cycle. Right now, let us acquire a new skill. Today, we will acquire the knowledge necessary to develop a new program. Therefore, we will develop a 50-number program that we have recently implemented. Therefore, activate the Edit key. This key will be acquired. However, it is important to recall this information for the time being. To create a new program, you must select the Edit key. Following this, you are required to compose all 0050. As you are aware, this is the designation of our program file. Subsequently, click this key. View our program in its entirety. This window is empty, as you can observe, as we are going to develop a new program in this location. Therefore, I will be composing a compact program. You are simply experiencing a cycle. There is no need for concern regarding the content of this document. As we progress through our syllabus, you will gradually comprehend all of it. However, for the time being, please observe the manner in which I am composing this program. You are only required to observe the key, its intended use, and the window that is opened as a result of striking the key in this subject. Only you are required to observe this program. Therefore, our program is now prepared. Immediately following that.

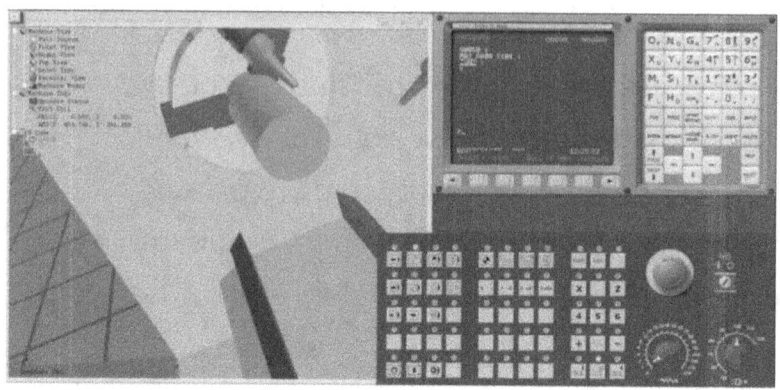

Enter the auto mode by pressing the key, and then initiate the cycle. Observe? The operation of our initiative is ongoing. Therefore, I trust that you have comprehended the process of developing a new program. Therefore, this pertains exclusively to the program key. Now, let us determine whether I have correctly stated the key. Therefore, this is the opposite key of safety that is employed to correct job sites. That is, we implemented limits or variant tolerance in order to preserve the extent of the task. We employ the opposite key. Alternatively, it is possible to assert that all tasks associated with slides are executed in the opposite direction. Therefore, there are three categories of antagonists. The initial example is starkly opposed. Second is geometry. The third factor is labor offset. Offset is employed to rectify ongoing projects on job sites. This implies that if you are engaged in production, we will be unavailable, resulting in the acquisition of additional sites that are necessary. Therefore, in order to eliminate the surplus material, we must induce some disturbance. Therefore, the instrument will also remove the excess material. Additionally, we obtain the sites that we require. So, these types of offsets are provided in "We are offsets." In our offset procedure topic, we will examine the proper method for providing offset and provide thorough details. However, for the time being, please examine the contents. The subsequent offset is the geometry offset. It is employed for the establishment of new jobs and the development of existing ones. This implies that if you intend to create a new position, you must employ this offset. Subsequently, employ geometry offset for the purpose of developing new positions. Additionally, if you are engaged in production at that time, you will offset to the rear. Acknowledged. Therefore, there are instances in which it is necessary to assign identical offsets to all instruments. Therefore, rather than specifying an offset for each tool, specify an offset on the work offset page to ensure that it is applied to all tools. All right. Therefore, these are the three offsets. So that you may gain a more comprehensive understanding of offsets, we will now demonstrate these offsets through the use of projects. Therefore, upon pressing the offset key, this window will be initiated. Within the window, hit the offset key. Additionally, upon selecting offset, you will be presented with two alternatives: geometry and wire. The two offset pages are virtually identical. As you are aware, geometry is employed in the development of novel jobs. Additionally, we are currently engaged in production labor. As a result, we are unhappy. Contains two timber pieces, x Z r and t. Therefore, x represents the diameter, Z represents the length or distance, R represents the nodes radius of the insert, and t represents the tool type. This serves as an introduction to offsets. We shall acquire comprehensive knowledge regarding each of our forthcoming subjects. Presently, we shall examine the manner in which to provoke discomfort. Therefore, specify the offset value that we wish to provide. Subsequently, select "plus input." Please bear in mind that you must select the "plus" input. If you select the input button, the tool will only accept the new upset and delete the existing offset. However, we are opposed to that. We wish to incorporate our new offset into the existing offset. Therefore, it is preferable to select the "plus" input key rather than the "input" Add input as your offset value to the existing value, and input will supplant the existing value with the new value.

WEAR	GEOMETRY
It is used for job size correction of ongoing job	It is used for job setting and new job development.

WORK: Work offset is for giving offset to all the tools at a time. That is if we give offset in work mode, all tools gets affected.

Therefore, it is crucial to consistently select the input key "plus" after providing an offset. Additionally, our third offset is Virgo's distress. Therefore, the offset specified on this page will be applied to all of the instruments in this simulator. I possess four credentials; however, this is not the case with your CNC machine. Your CNC machine will have a single x and z axis. If your CNC machine also possesses these four identifiers, please enter the x and z values in this field. Additionally, the code that we are employing to provide an offset must be incorporated into our primary program. However, the new control interface contains only one x and z to indicate the task offset. Therefore, it is unnecessary to incorporate that code into our primary program. Therefore, if you possess four codes, select any code for your work offset. Provide the values of x and z in that location. Additionally, ensure that this code is incorporated into our primary program. In our offset topic, we will cover every aspect in detail. Therefore, there is no need for concern; this serves as an introduction. All right. Therefore, we will proceed with the subsequent section.

CONTROL PANEL (PART 2)

Therefore, we should proceed. The subsequent subject is diagnosis or system. Therefore, these are employed for the purpose of maintenance. Therefore, refrain from pressing this key, as it contains the entire CNC configuration. The subsequent key is a custom graph key. Graphical representations of the program's execution are presented here. This is not noteworthy. The tool's graphic functionality is exclusively visible to you. Therefore, to access it, simply tap this key. During the execution of any operation by your CNC. The subsequent key is the message key. With the assistance of this key. The message is visible. Therefore, CNC will notify you via message if you have made an error. Therefore, to access that message, select the message key. Observe the error you have committed. Resolve the issue and select the reset button. Therefore, let us take a look at this. I am going to make an error. Therefore, I am receiving this message in this simulator. In this manner. However, this message will be displayed in your CNC machine. All right. Therefore, resolve this issue and select the reset button. Therefore, we should proceed. Additional selection keys comprise our subsequent keys. Therefore, there are specific modes in a CNC machine. Therefore, these controls are employed to select that mode. Additionally, we will execute certain operations with the assistance of these modes. Therefore, we will examine them individually. Therefore, our initial mode is default. Therefore, the machine must be in auto mode whenever you wish to initiate or execute a program. The program will not function without it. Additionally, CNC will transmit a message to you. Therefore, in order to execute your application, you must activate auto mode. The subsequent mode is edit mode. As you may recall, we activated edit mode when we developed a new program at that time. It was subsequent to this that we developed the program. Therefore, in order to generate a new program, it is necessary to select the "edit" mode. Additionally, any actions taken in this mode are permanently stored in your CNC machine. Therefore, any modifications made to your program are stored in your CNC machine. The subsequent mode is MDI manual data input. Therefore, if you require any modifications for a transient purpose, you must implement them in MDI. Therefore, we will examine this matter through the use of project. Therefore, this is MDI. Therefore, I wish to rotate the spindle. Therefore, we should rotate it. We have three codes for that purpose: M0. Subsequently, we shall examine this course. However, for the time being, please examine the contents. I press the "cycle start" button, and the spindle rotates, and the code vanishes. Therefore, this demonstrates that this mode is intended for a transient purpose. Reference is the subsequent mode. Remember, in our initial discussion, I advised you to perform homing when you initiate your CNC machine. Therefore, this is a reference to the homing function of this turret. Move to the position of reference. Therefore, we will examine the process of localization with the assistance of a project. Therefore, this is our fortification. Initially, I will set this correct to ensure that you have a comprehensive understanding of homing. All right. Therefore, select the reference key, followed by the x and Z keys. Additionally, our turret is being relocated to its reference position. The subsequent mode is pace mode. Therefore, upon activating this mode or key, the turret may be rotated or moved. Therefore, you will inquire as to why it is necessary to rotate the turret. Consequently, our insert is frequently damaged or worn out. Therefore, in order to modify that insert tool, it must be present. Then, you will be able to effortlessly modify that embed. In order to have the tool in front of you, we must first select motion mode and then press the turret key. This will ensure that the tool you require is in front of you. To modify that, insert. Then, modify the insert. This cycle mode is also employed for instrument movement. Therefore, we will examine this exercise mode with the assistance of a project. So, initially, select the exercise mode. Subsequently, I shall execute tool movements.

Observe that our instrument is in motion. All right. The subsequent mode is manage mode. Therefore, tool movement can also be performed in handle mode. In this location, we can regulate the pace of the tool's movement by tapping these children. Initially, select the "multiply 1000" button to initiate tool movement.

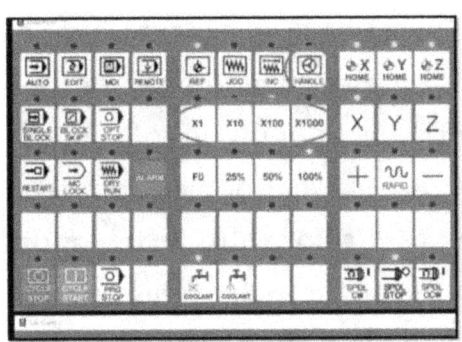

HANDLE

- Used for tool movement manually.

Additionally, for sluggish movement, select either multiply ten or multiply one. The instrument must be touched to jog press, depending on the situation in the configuration. Consequently, we will adjust the tool's movement pace at that moment. These keys will be implemented. Therefore, this is the appearance of our handle. The following keys are program control children. Therefore, we can manage the initiative with the assistance of these children. Therefore, our initial key is a single unit of b and k. This key is employed to execute the program line by line or block by block. This implies that the machine will read the first line, execute it, and then halt again. Upon pressing the cycle start button, the machine will read the second line or block. Run it and then halt. Therefore, we will examine this matter through the use of project. Therefore, I have chosen a solitary unit. Therefore, observe the machine as it operates a single block and then ceases. Subsequently, the cycle will commence once more. Read a single line, execute it, and then cease. Therefore, this is employed when creating a new position. In that scenario, it is imperative to determine whether the program is functioning properly or if there are any errors. In order to prevent instrument collisions or accidents. With the assistance of this key. Therefore, when developing a new task, it is imperative to utilize single block mode. The subsequent essential is the deletion of the construct block. Build mode is available for the purpose of skipping any block or line in the program. If you do not wish to include any line or block, simply add a slash before the line and select "Build mode." Therefore, we will examine this in greater detail with the assistance of a project. Therefore, this is the key to our build mode. Additionally, this slash must be applied prior to that line in order to bypass it. Therefore, we will determine whether or not the line skips. Therefore, the C sentence is skipped. Therefore, it is operational. Optional stop is the subsequent key. As we have observed in the single block machine, it halts after a single line or block. The machine will halt after a specific operation in the optional stop mode. That is, the program contains a multitude of operations, including roughing, drilling, polishing, and grooving. Therefore, to terminate the machine following a specific operation, execute the M01 code and then select the optimal option. Are you receiving the distinction in single block machine pauses for each block? Additionally, in the optional halt mode, the machine terminates subsequent to a specific operation. Therefore, we will examine this matter through the use of project. Therefore, as you can observe, I implemented the M01 code following the initial operation and selected the discretionary halt mode. Please observe the machine's halt. Now, let us observe the outcome if I disable OPD mode. Therefore, the machine continues to operate even after the M01 code is executed. The second instrument is poised to execute its operation. Therefore, it is necessary to select the operating mode in conjunction with M01. The machine will cease to operate only after this point. Please bear this in mind. Operational mode must be activated in conjunction with the M01 code. The subsequent essential is the trial run, which involves the machine being operated without any actual cutting. This is done to observe two distinct moments. This key is employed in this context. Every movement is rapid, and cutting pace is not taken into account. Therefore, this mode is not frequently employed. It is possible to observe complete movement without any actual material cutting. All right. The subsequent key is the machine loc block machine moments or tool moments. Additionally, this key is of limited utility. Please observe the accompanying project regarding this key. So, I have chosen machine lock mode. And when I select the "cycle start" button, I observe that only the spindle is in motion. Every other movement is obstructed. The following keys are editing gates. Therefore, these controls are employed in the process of program writing and modifying. Therefore, the initial key is the "insert" key, which is employed to enter written code into the CNC. Therefore, we will observe this key in a project. I will enter a line of code here. Additionally, when I press the "insert" key. Observe? Our code is generated by a CNC machine. Therefore, this is the application of the insert function. Here, I will compose an additional piece of code. Next, select "insert" and then " Therefore, the subsequent key is the delete key. In the event that you inadvertently compose an incorrect code and wish to eliminate it, include the time at which the key was utilized. Therefore, simply observe the project. I will be deleting the G01 code in this location. Therefore, select the code in question and strike the delete key. That concludes the matter. Therefore, the subsequent key will be "turn." If any written code is incorrect and you wish to replace it with another code, you may employ the "alter" key. Therefore, we should observe this in a project. Right here. The process of modifying g zero and coding with G00 is as follows. Initially, choose that code. Then, I modify the code. This is the method by which we modify. Therefore, the subsequent key is the shift key. So, as you can observe, a single key contains two alphabets. To print the second alphabet of that key, press the shift key and then the key in question. We will now observe this through the use of project. Therefore, observe how I utilize the shift key to select u and W. First, hit the shift key, and then press the u key. And the same for w, alright. Therefore, in Fanuc, we are required to press the shift key only once before selecting any desired key. However, in Siemens, it is necessary to press and maintain the shift key. This is followed by the pressing of an additional key that is intended for printing. Therefore, there is a minor distinction between these two control panels. In Fanuc, a single stroke is sufficient. The shift key must be pressed and held in Siemens. We will proceed to the next key, which is the cancel key. If we make an error while composing a program and wish to eliminate the code, we utilize the "cancel" key. Additionally, the distinction is apparent. The delete key is employed to delete any code that has already been written or inserted, as well as to cancel. The key is employed to eliminate code that is not inserted during the programming process. Okay, let's examine this with the assistance of a project. I will make an error while composing the program. Additionally, I will employ the "cancel" key to rectify that oversight. Okay, so if you make an error while composing, simply press the "cancel" key. Additionally, the delete key may be employed to modify previously entered code. The input key is the subsequent key. This key is employed to induce distress. Therefore, we will examine this essential in the contrary subject. Therefore, we should conduct a revision. What we have acquired. Therefore, these are all alphabets, and these are the numerals. Here, we have our block key and the EOB key widget. Which is employed to conclude the block. Therefore, the machine will recognize that the block has concluded when you select this key following any code. Right here. Afterward, we acquired the knowledge of the position key, which is utilized to determine the position of the tool program key in order to assess the program's ability to cause an upheaval. We have three kinds of offsets. We have a work offset and geometry. Then, we observed the shift key, which is employed to select the second alphabet of the key. The input key is utilized to provide an offset, which we will learn about later. The cancel key is used to terminate any code that is being written in the program. A system key that is employed for maintenance purposes. To view the message, press the message key. A custom graph is available to view a graphical representation of the program in progress. I will proceed to modify any code with the addition of a new code in order to incorporate it into the CNC. Press the delete key to eliminate the code that has been inserted. There is an additional page.

Return to the previous page. These are employed to elevate the reticle. Descend. Left, right. This is the help key, which is used to obtain assistance. The reset key is used to halt the machine in an emergency and to resume the program. Okay, let's examine the contents of our primary panel. Therefore, we are in automatic mode. We must select this code to initiate the program edit key, which is necessary for program modification and the creation of new programs. In this configuration, these modifications are permanant. MDI is the subsequent step. These modifications are transient. This key will be employed to rotate the spindle for a brief period. All CNC machines are equipped with remote keys. Presently, this key has been eliminated. This key was employed to operate the machine via remote control. The subsequent key is a reference that is utilized in the home internet or tool motion mode to rotate the turret and perform tool movements. Handle and ANC. Both are employed to facilitate the movement of tools. ANC has as of now been abolished. These are intended to increase the efficiency of instrument movements. For rapid tool movement, multiply by 1000, and multiply by 100. Divide by ten. Slow movements will result from multiplying by one. These pertain to the movements of the x and z tools. This machine operates on a single cylinder. brick by brick. Block. To halt the machine after a specific operation, use the slash OPD to proceed to the next block. Restarting a tool is equivalent to resetting a machine safeguard to prevent tool movement. In a dry run, the machine is operated without any actual cutting of material, and an alarm or message is displayed to determine the status of the alarm. This reading is expressed as a percentage. When the instrument is given any weight in the program, the input is considered 100%. And if we wish to decrease that, you may reduce it to 50% or 25% as you wish. This is the cycle stop to stop cycle cycle start program stop to stop program coolant or coolant of as we know we have rotated spindle in MDI with N03 code. Therefore, these controls can also be used to rotate the spindle. Therefore, the entire control panel has been viewed with the assistance of a project. I trust that you comprehend all of the keys and have a clear understanding of the z variable. Therefore, this is exceedingly advantageous for operational purposes. So, we have now concluded our third topic.

OFFSET PROCEDURE

The fourth topic we will discuss is the opposite procedure. In this section, we will examine the process of calculating the offset value and providing it to the machine. As we are aware, offsets are classified into three categories. Therefore, our initial offset is the we are offset, which is employed during production to insert we are. Therefore, we will investigate the process of determining the offset value. Therefore, this is our workpiece, and the desired number of sides after machining is 28. However, the fact that we are receiving 28.02 as a result of the insert indicates that there is an excess of material present. Therefore, in order to determine the excess material, we must subtract the desired sides from the sides that we receive. This implies that we must subtract 28 from 28.02. Therefore, we end up with an excess of 0.02 material. Additionally, this is the offset value. This value will be expressed in x as the additional material of diameter. Therefore, we will assign an offset of 0.02 to x. Currently, the instrument selection process is underway. Therefore, we are. The finishing tool is the primary recipient of offsets, as it is responsible for the final cut. Therefore, the excess material will be eliminated. You will now inquire as to how we determine which finishing instrument to use as the lead. To accomplish this, it is necessary to inspect the interior of your machine and determine which instrument is responsible for the refining process. This is the method by which you will determine the number of your finishing instrument. Alternatively, the tool numbers are inscribed on the turret if you observe it closely. Therefore, the finishing tool number can be obtained by examining the tool number of the finishing tool. So, we now have a value of 0.02 for x. Additionally, this offset will be applied to the finishing tool.

How to calculate OFFSET value?

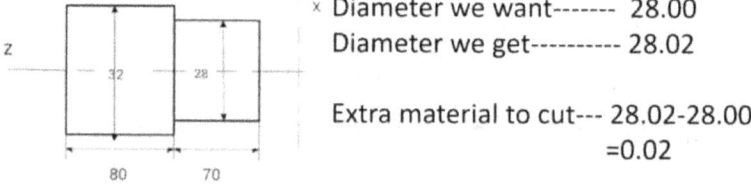

x Diameter we want------- 28.00
Diameter we get---------- 28.02

Extra material to cut--- 28.02-28.00
=0.02

It is now necessary to determine whether to provide this offset in the positive or negative direction. To that end, we will examine the following. The additional material is evident. Therefore, it will be necessary to remove the excess material. As you are aware, the graph of x and z is negative in x and positive in x. In the same way, the right side of Z is positive, while the left side is negative.

Therefore, it is evident that our instrument is descending in x. That is the opposite of plus. Therefore, we will relinquish the set at x -0.02 to the finishing tool. Therefore, we have acquired the ability to calculate offset. Therefore, this pertains to the offset of x. Let us determine the offset of any material that is present in Z. Therefore, the desired length is 70.50 meters, which is equivalent to M. Therefore, the additional material is 0.5 meters. Therefore, the value obtained is the length value. Therefore, we will relinquish the set in Z and select a tool. Who will be responsible for trimming the excess material? All right. Therefore, we have two numbers, z, and a value of 0.5. Now, we must determine whether the value of 0.5 is positive or negative. Therefore, as we are aware, the right side of the x and z graph is positive when two layers are moving. Additionally, it is negative on the left side. Our instrument is currently in motion to the left. Therefore, it is negative. Therefore, it is necessary to specify an offset of -0.5 in Z. Therefore, we have acquired the ability to calculate offset in both x and

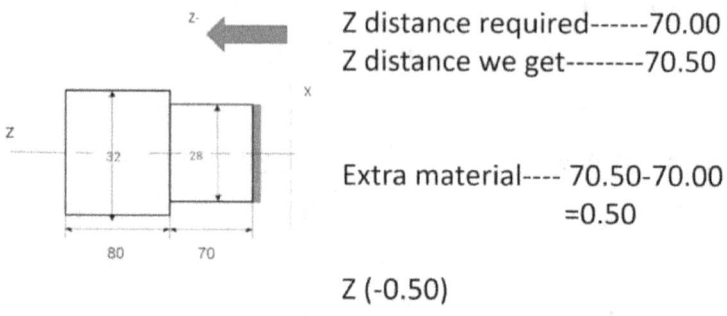

Z distance required------70.00
Z distance we get--------70.50

Extra material---- 70.50-70.00
=0.50

Z (-0.50)

Now, we will examine the process of assigning this offset in the control panel. Therefore, this is the methodology. Therefore, our initial objective is to examine the procedure for final control in order to provide an offset. So, initially, press the offset button. Then, select the offset option again in the offset button. Following this, you will be presented with two alternatives. Select the "We are" option, as we will be providing an offset in V. Subsequently, select the tool number to provide an offset. The finishing implement is the most frequently chosen. Subsequently, determine the X or Z value that will provide an offset in length or diameter. So, you must first choose either x or Z, and then input the offset value you wish to provide. Therefore, we have acquired the ability to compute this offset value. To do so, input the offset value and then select the plus input key. As previously demonstrated, the input key must be pressed in addition to the input key. Additionally, the input key will be included. With the existing offset, you are already offset.

And that is precisely what we desire. Therefore, consistently utilize the input key "plus." So this is the process we follow.

Procedure for Fanuc control

1. Press Offset button
2. Press offset option
3. Press wear option
4. Select tool number to give offset (mostly finishing tool)
5. Select X or Z value
6. Enter offset value you want to give
7. Press +input key

Now, let us observe this procedure through the use of a project. So, press offset once more, and we will now select the total number. I will employ the fourth number tool as my final instrument in this instance. Then, you press the plus input key and agitate -0.02 and x. That concludes the matter. Please recall the input key "plus." Then, just select "auto mode" and begin the cycle. Therefore, observe how straightforward it is. Therefore, this is the manner in which we must contribute. There is considerable distress among us. Therefore, the issue at hand is the ultimate control. However, there is a minor modification to the offset calculation in Siemens control.

Therefore, Siemens control encompasses both the original Siemens control and the new Siemens control. Therefore, we will examine the procedure for all Siemens controls. So, initially, select the offset key, and then select the tool number. This is the number of the finishing instrument that is most frequently employed for offset. Subsequently, provide the offset value. In this case, all calculations must be performed manually using the outdated Siemens new symbol control. Perform each of these calculations. However, it is necessary to perform all of these calculations for all Simmonds. Your comprehension will be significantly enhanced by viewing a project. Therefore, we must input the value that is required in the offset field, then press the input or calculate button, and finally press the "accept" button. Therefore, in order to facilitate your comprehension, we will demonstrate this procedure through project. Therefore, select the tool by pressing the offset key. At this point, I wish to provide an offset for the initial tool. As you can observe, the -0.5 offset is already present in X. Additionally, if I wish to elevate the instrument, this entails an additional 0.2. Therefore, I must combine the 0.2 with the -0.5. Therefore, we are required to perform all of these calculations manually, as Siemens is responsible for the entire process. It is imperative that we perform this calculation. Therefore, I must combine this value with -0.5 and add 0.2. So, the result of the calculation was -0.3. Substitute that value into the variable x. Then, press the input key. This is the method by which we must cause dissatisfaction throughout Siemens. However, Siemens has implemented a novel management system. This calculation is performed by Siemens. Therefore, let us examine its methodology.

Procedure for Siemens control

Old Siemens Control
1. Select Offset key
2. Select tool number
3. Give offset
4. Input/calculate
5. Accept

New Siemens Control
1. Select Offset key
2. Select tool number
3. Press =
4. Give offset in calculator
5. Press input/calculate
6. Press Accept

Therefore, in the new Siemens control, select the offset key and then the tool number. Furthermore, you must select the X equal to sign that is located on your Siemens control interface. Therefore, select the key that corresponds to the corresponding button. Additionally, upon selecting that key, you will be presented with a calculator. In that calculator, the value of -0.5 is already present. We are now required to input our plus 0.2 value into the calculator. After providing our value to the calculator, select "input" or "calculate" to ensure that the

CNC machine automatically calculates its object value. Afterward, select "accept." This is a minor distinction among all Siemens controls. You do not receive a calculator; however, a calculator is provided in the Miele Siemens control system for calculation purposes. Therefore, I will execute this Siemens control procedure once more. First, we must choose the offset key. Subsequently, we must select the tool number and then press the X equals sign. Therefore, upon selecting the equals sign, we are presented with a calculator. The calculator is already present. The offset will be indicated. Additionally, we must include our offset value, which is 0.2, in that offset. Therefore, input the value of +0.2 into your calculator and then select the input or calculate icon. Therefore, the calculator will perform the calculation and subsequently select the z key. Subsequently observe. So, this is the manner in which anger is communicated in the new Siemens control system. Therefore, this is a minor distinction between the total Siemens control and the novel Siemens control that we have examined. We will now examine the labor compensation. Therefore, the offset specified here will be applied to all instruments. Therefore, this is the appearance of our offset page. Simply enter the offset value in x or Z and select the plus input. That concludes the matter. It is implemented automatically. Our third offset is geometry offset, which we will explore in the setting topic, as it is employed for the purpose of setting. In this topic, we have acquired the ability to calculate offset values using the plus and minus signs. All right. So, we have now concluded our fourth topic.

CAREER GUIDANCE 1

Therefore, individuals who are unfamiliar with this discipline may encounter some challenges. You may perceive that I am merely loading and discharging the tasks, and you may also believe that this is the case. Also, how can I become a programmer with this? Therefore, if you are experiencing all of these inquiries, it is advisable to exhale and exercise patient as this process is underway. Or, you may not have a habit of performing rigorous work, which may result in early fatigue and numerous challenges. However, you are not required to give up and must continue with your operational responsibilities. You may encounter numerous challenges during the initial stages, and there is a possibility that you will be compensated at a low rate. However, this is merely the beginning, and if you undertake any additional work, you will encounter additional challenges during the initial stages. Therefore, do not surrender and persist in your endeavors with additional diligence. Therefore, you have reviewed the initial four topics and are now required to complete one task. You are now required to view seven topics that pertain to measuring instruments. Therefore, the initial four topics and this seventh topic.

So, in total, there are five topics. Therefore, these five subjects are specifically designed for the purpose of CNC operation. Therefore, if you are a novice in this discipline, it is sufficient to concentrate on these five subjects. Additionally, you are required to view these five projects on a continuous basis for the next three months.

7. MEASURING INSTRUMENT

TOTAL=5

3 MONTHS

You must view these projects as frequently as possible. Watch these five projects and acquire the full knowledge that we have provided in them. Additionally, it is imperative that you implement this knowledge on your CNC machine. This obligation pertains solely to the subsequent three months. Refrain from proceeding. Watch these five projects and implement all of the knowledge to your CNC machine. This is your foundation, and it must be robust. Only then will you not encounter any challenges in your pursuit of becoming a programmer. Your base will remain vulnerable and you will encounter numerous challenges if you attempt to accelerate your progress. To become a programmer, as your foundation was inadequate. This is the reason why you must view these five projects in a continuous manner for the next three months and implement all of the knowledge to your CNC machine and the professionals who have been working in this field for a long time. They have the option to view additional projects; however, those who are new to this field are required to adhere to this policy. For the next three months, refrain from engaging in any form of dishonesty. View five projects and implement all the information on your CNC machine. Your foundation will be fortified as you attain perfection in these five subjects. Therefore, you must exert considerable effort over the next three months. It is imperative that you dedicate significant effort to these five subjects in order to avoid any future challenges. Consequently, you are required to concentrate solely on acquiring knowledge for the next year. Refrain from contemplating salary. Concentrate solely on information. It is imperative that you acquire an increasing amount of knowledge and apply it. Observe and determine how to augment your salary. What is the method by which you will increase your salary? This is the individual we will observe in our final career guidance project. However, at this time, it is imperative that you concentrate solely on acquiring knowledge. Therefore, observe these five subjects and dedicate the subsequent three months to their advancement. Then, what are the subsequent subjects? The only way to become a proficient CNC programmer in a short amount of time is to adhere to my path. What strategies will you employ to secure a significantly higher salary within the next one to two years? That is also something that we will inform you of. However, for the next three months, you will be required to devote a significant amount of effort to our five subjects. Additionally, it is imperative that you possess a comprehensive understanding of these five subjects. You are required to proceed further only after that. Therefore, engage in diligent task and only thereafter view additional projects.

JOB SETTING

Therefore, our subsequent subject matter is employment. Therefore, we must establish a setting for each new job development. Therefore, there are two configurations in this work. I will begin with the work environment. The second factor is the job environment. Subsequently, we must implement task configuration. Therefore, we will examine the process of task arranging. Therefore, in your environment, it is necessary to initially supplant all existing positions with new ones. Consequently, we are compelled to perform tedious tasks following our job transition. Therefore, we will examine each item individually. To eliminate all tasks from the job list. With the assistance of a key, we must eliminate its morality. However, attempting to remove these bolts directly will result in the spindle rotating, rendering it impossible to apply the complete force necessary to remove the jobs. To accomplish this, it is necessary to initially secure the spindle. Additionally, we possess the M 19 code for this purpose. Write 19 lines of code by selecting MDI more. And then, press the "cycle start" key. Therefore, the spindle will become locked. Additionally, we will observe this through the use of project. However, for the time being, please examine the contents. Therefore, after the spindle is locked, it will be possible to remove tasks and subsequently install new ones. Therefore, it is recommended that new tasks be loaded with approximation and minimally compressed. The fasteners are not completely secured. As a result of our intention to modify these positions. After that, utilize it as a tool in close proximity to your workplace. Additionally, by employing the tool to verify the distance of these tasks, we can determine whether they have been imported correctly. Therefore, let us examine each of these points in greater detail. Therefore, our initial objective is to secure the spindle in preparation for the eradication of the work. Therefore, observe the process of securing the spindle. To begin, select MDI. Then, select the rightmost M 19. Afterward, the period concludes.

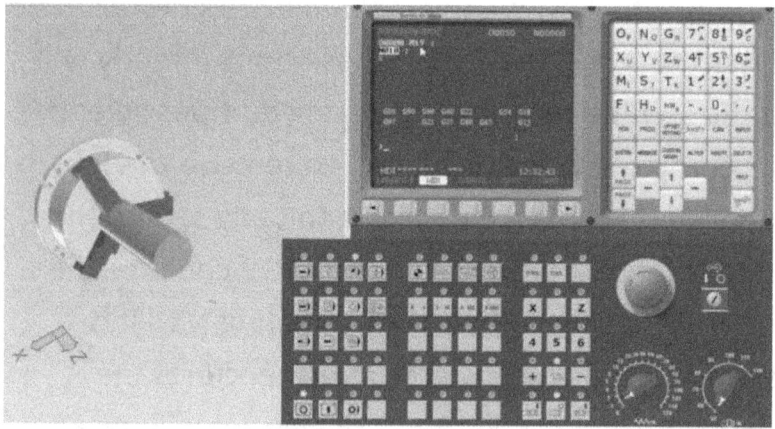

Subsequently, proceed with the insert and subsequent cycle. Therefore, our spindle is secured. Subsequently, we shall eliminate all positions. We are now required to upload new tasks. Additionally, it is necessary to determine the position for the individual in question. Jobs of this nature that are capable of maintaining their position. Our jobs are not adequately matched to the position, as evidenced by the initial image. In the second image, our tasks are appropriately matched to the role. So, opt for these types of positions. This knowledge is exclusively acquired through experience. All right. Therefore, the more you engage in this subject, the more you will acquire this knowledge. Therefore, we have made the following selections of new positions. Therefore, in order to place these tasks onto the check, we have serrations behind the jobs. These are the serrations, as is evident. The same type of serrations are also present in the workplace. Consequently, it is advisable to accommodate this new position for the most part. In the project, it is evident that the suggestion is to mildly tighten the task. So in this manner, load all three tasks in order to achieve the following appearance of your work. Following this, it is necessary to bring an IED instrument into close proximity. Worked in positions similar to this. Acquire a compact IED instrument. Therefore, it will be placed between these tasks. Afterward, it is necessary to recall this distance. I will then rotate tasks to verify the distance for the subsequent job. This is a requirement for each of the three positions. If the distance is not uniform across all tasks, you must modify them by moving them forward or backward. Then, verify the distance for each assignment once more. Therefore, you must repeat this process until the distance is consistent across all tasks. Your employment selection is accurate if this distance is consistent across all positions. After that, return the tool to its original location and then thoroughly connect these tasks. This is the method by which you must establish your employment. Following this, we are required to complete the task. Loading. Therefore, this task is completed to ensure that the work is aligned correctly and accurately, as well as to obtain an accurate measurement. Due to this, we implement work boarding. This is the visual representation of job encoding, as you can observe. You must only make a slight reduction in the x and z directions. Therefore, we will investigate the process. Therefore, it is advisable to establish a fictional business in the interim between employment opportunities. This artificial piece must be significantly smaller than our workpiece. Rotate the spindle and then contact the boring bar twice in the x direction. After that, open the position window.

Procedure

1. Hold dumy piece in between jaws
2. Rotate spindle
3. Touch boring bar to jaws in X direction
4. Open **position** window ⇒ **Relative** ⇒ Write **X** ⇒ **Origin**
5. Touch boring bar to jaws in Z direction
6. Write **Z** ⇒ **Origin**

In the position window, hit the relative key in the relative right x direction, and then press the origin key. Afterward, you must touch the uninteresting bar to the positions in the Z direction. After that, enter z in the position window and select the origin button to ensure that both x and z are zero. Therefore, this is the appearance of our mock-up base. Our prototype piece is significantly larger in this instance, as we anticipate a workpiece of considerable size. Therefore, it is imperative to maintain that your artificial base is significantly smaller than your workpiece. Therefore, this artificial portion should be retained in between tasks of this nature. All right. After that, rotate the spindle with an m-zero record and then take a dull bar in between these tasks. And then, gently touch the tasks in the x-axis with this boring bar. Then, select the position key, followed by the relative key, and finally the right x key. Write x if your control interface contains x. If your control interface includes the letter "u," select "u" and then "origin." In the same way, lighten the contact of the boring bar to the adjudicator in the direction of Z. Then, select "position relative" and "right Z" or "w" depending on the controls on your control panel. Then, select the origin button to ensure that your x and z values are zero at this juncture. Therefore, in order to facilitate your comprehension, we will demonstrate this through project. So, after touching the tool to evaluate in the X direction, select position, then relative, and finally right you, as we have you here. Then, the point of origin. Observe that our U becomes zero at that juncture, alright. In the Z direction, contact the mandible with the tool, followed by the right w and the origin. Therefore, we have established both nodes at zero zero. We will now commence the process of task loading. So, how do you slash a judge? So, initially, we must remove the instrument from the mandible. Next, move the tool in the X direction until it reaches zero. Therefore, we have established these x and z values at zero zero. Move the tool upwards until you achieve the desired level of incision. I am proposing to reduce the amount by nearly 0.2 million. Therefore, I will increase the tool's value to 0.2. Subsequently, we must execute a 0.1 m reduction in the z-direction. The instrument must be relocated in the z direction by up to -0.1 m. Then, we must remove the tool in the x direction and again remove it from the judgment.

Start jaw cutting

Therefore, we have implemented the initial reduction of all tasks. This procedure must be executed twice. Therefore, we should examine the second incision. In the second incision, advance the tool in the x-axis by 0.2. Given that we have already implemented this reduction. We are now required to make another 0.2 MB reduction. Move the instrument 0.2 NM upwards in the x-axis. Therefore, the total length of the x-axis is 0.4 meters. Then, we must make a trim in the z direction up to -0.2. Given that we have already implemented a 0.1 reduction in the z-direction. Again, we will reduce the amount by 0.21 million. Therefore, the total reduction is -0.2 meters. We must proceed in the direction of Z. Afterward, we will execute an undercut in this location. Therefore, this is the appearance of an undercut. This is performed to ensure that the work is clamped securely and to prevent any misalignment. Therefore, it is imperative that you implement this undercut whenever you engage in tedious activities. Therefore, we will elevate the instrument marginally upwards from Z -0.2. And then, move away from the adjudicator by taking the x-axis downward. Boring is executed in this manner. Therefore, I will execute this procedure on the control panel to facilitate your comprehension. Therefore, we have already established U and W as the origins in this case. Additionally, our uninteresting bar is situated away from the judge. Therefore, I will extend the tool to 0.2 in the x-axis in order to make a nearly 0.2 m incision in the x-direction. So, I have nearly attained 0.2 meters. Subsequently, I will reduce the Z-axis by -0.1. This is me, accelerating its velocity. However, you individuals proceed at a deliberate pace and with precision, as we are in the process of cutting material. Therefore, it is necessary to proceed cautiously. Observe that we have acquired a value of -0.21. Then, we must move the instrument downward in the X direction. Below zero. Then, remove the instrument from the judge. So, we have finished our initial edit. Once more, we must elevate the instrument to a height of 0.4 m in the X coordinate. Given that we are planning to perform another 0.2 m reduction. Therefore, it is imperative that you execute the task correctly.

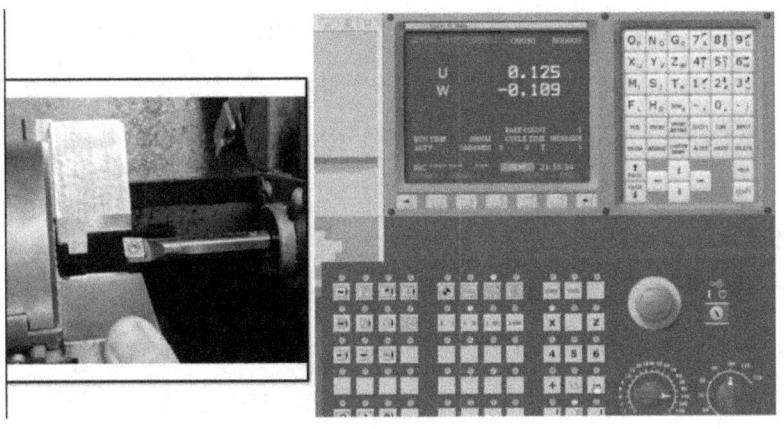

Due to my utilization of this simulator. Therefore, I am incapable of performing the task in a manner that is satisfactory. However, you individuals execute it correctly. All right. So, we have reached nearly 0.4 M in the X direction. The instrument must be moved in the z direction until it reaches -0.2. Therefore, I will conclude at this point. We are now going to conduct an incision in this location. Therefore, elevate the instrument slightly in the x-direction. Subsequently, downhill. Below zero. Then, remove the instrument from the judge. At this point, we have acquired the knowledge necessary to load, evaluate, and perform job loading with an undercut. All right. We will now examine the job context. Therefore, what is the definition of employment setting? Therefore, it is necessary to determine the offset values for each instrument in terms of x and z. Additionally, the geometry offset page is included. This is a work environment. That is, you must obtain the x and z offset values for each instrument and subsequently enter these values into the geometry offset page. We will now examine the process for achieving this. Therefore, the instrument is initially applied to your face while the spindle is operating. After that, press the offset key. and press offset once more in that offset. Then, you must select geometry. And in geometry, select the tool number. Therefore, you must select the tool number that you most recently touched on the mandible face. After that, select Z and then enter z 0.0. Then, select "major." Therefore, this pertains to the z offset. Presently, we shall investigate the offset of x. The same instrument is used to measure the diameter when the spindle is activated, and then x is selected. Additionally, it is necessary to specify the diameter of the project. Write z 0.0 in the z offset. However, the diameter of the task must be specified in x. And subsequent to that, press the major. We will now demonstrate this through the use of a project to facilitate your comprehension. Initially, it is necessary to apply the instrument to the mandible face. Therefore, it is imperative that we execute this task in a manner that is both precise and appropriate. This selection is essential, as it determines the entire assignment size.

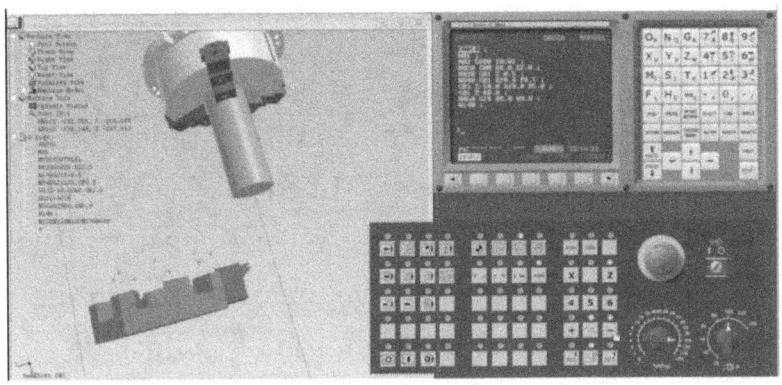

Therefore, approach the task with the utmost concentration and endeavor to establish yourself as an accurate individual. Therefore, we will rotate the spindle at a speed of 800 r.p.m. The more precise the parameters are, the greater the accuracy of your work. Occasionally, it is necessary to create a single position. At that period. Accuracy is necessary. Therefore, it is preferable to establish oneself as a precise individual. Subsequently. Touching. A tool is applied to the mandible. Pressing the face of Z. Please press the key again. Offset. Then, geometry. Then, place the cursor on Z. Additionally, Z 0.0 is located here. Then, measurements are taken. Repeat the process for your confirmation. We must now apply the same instrument to the diameter of the task. I will elucidate this setting by utilizing an offset for a single instrument. However, it is necessary to account for the offset of all the instruments that will be utilized in your program. Consequently, the diameter of the task should be specified after the tool is touched to the right x. I am employing a diameter of 80 meters in this instance. So, right x 80.0. Afterward, we must input the radius of the inset of the nodes that we are using in R to measure. This is the type of instrument that we will be observing in the near future. Therefore, this is the method by which you must conduct your job arrangement. Therefore, let us examine the type of instrument. Therefore, the final column of the geometry page pertains to the variety of instrument. Consequently, the following are a few numbers. Arrange in a specific order. It is imperative that you retain the sequence. This is the direction of insert trimming. Therefore, if our tool is cutting material in this direction, it is necessary to assign a value of one in the "tool type" column of the geometry page. This indicates that our insert is also cutting material in this direction. This zero-one instrument type is seldom employed. These are the subsequent steps. This cutting direction will be observed during the twisting of an IED. So, it is important to remember that zero two is for IED reversing. Thirdly, if the cutting direction of the Insert is of this nature, its tool type is zero three. This is evidently in the process of being turned. This implies that zero three is primarily employed for the purpose of reversing. Use zero for tool type zero five if you wish to perform only a reverse turn. It is intended for this particular trimming orientation. This is the utilization of a re-arrest instrument. It is evident that it is cutting through it in this manner. Zero six is employed when it is necessary to create a groove at that particular moment. The instrument type zero six is employed with the zero seven number. When we are required to perform phase grooving and zero eight when we already have a deep groove. So this pertains exclusively to the form of instrument. I have composed a message for you. The primary application of zero three is to indicate that the wheel has already begun to revolve. Additionally, zero two is principally employed for the purpose of returning I. Therefore, it is important to recall these two categories of instruments. Zero three is designated for returning, while zero two is designated for I returning. Therefore, if you recall these two instrument categories, you will have more than enough. Therefore, we have observed the entirety of the Fano setting. Therefore, we will observe the environment in three months. Jawboning and jaw alignment. It is identical. The sole distinction is that the OP is provided with a set in the geometry page. Initially, we should familiarize ourselves with the process. Therefore, the tool is initially applied to the mandible face while the spindle is in operation. Afterward, select "offset." Subsequently, we are required to select geometry in a disturbed state. Additionally, in geometry, select the tool number that is being utilized to configure the tool. After selecting two numbers, press the "major" tool key. Additionally, select the main Z key while using the measure tool. In major z, we must first set z to 0.0

and then set the extent of z. Therefore, this is the method by which we establish job positions in Z. The tool is now applied to the task diameter while the spindle is operational. Then, we must press the main x key. Additionally, it is necessary to specify the diameter of the project. Then, after inputting the diameter of the task, select the z-axis and the length x. Therefore, it is evident that there is a slight distinction between Siemens and Fanuc. Therefore, we will acquire this knowledge through the use of project. Select geometry by pressing the offset key. Here, I will select a single number instrument. Then, measure the tool's main x and right diameter. Right here. Then, establish the length x. Therefore, this pertains to x. Presently, does the instrument fulfill that function facially? Subsequently major z and write z 0.0. And then, the extent of z. That concludes the matter. Our configuration has been finalized. There is only a minor distinction between Siemens and Fanuc. Therefore, we have observed the settings in both control panels with the assistance of the project. Thus, we have concluded our discussion on the subject of setting.

TOOL SELECTION AND ITS NOMENCLATURE

Let's start with our sixth topic. So our sixth topic is tool selection. That means how to select tool that we use. So in this topic we are going to see insert selection external tool holder selection internal tool holder selection and insert failures and its remedies. That means we will see how to select insert external tool holder, internal tool holder and also we will see insert failures and its reasons. And how can we avoid insert failures. So let's start. So first it's insert selection. Here I have taken an insert for an example so that I can teach you these insert selection with the help of this example. So here we have ten MG 14 0408. These are total seven parameters. And each parameter have its meaning. So let's learn this in detail. So first is shape of insert. That means here shape of insert is taken at first parameter. Second is insert clearance angle. Third is tolerance grade. This key breaker and clamping. Fifth is size of insert sixth is thickness and seventh is node radius. So these are total seven parameters. So let's learn them one by one. So our first parameter is shape of insert. We have regular shapes and irregular shapes. So what is shape of insert. It's regular or irregular. Regular shapes means standard shapes like triangular, rectangular, square and irregular shapes means shapes that we get after some modification in regular shapes. Here D means triangular shape, means shape of that insert is triangular. Are means round. L means rectangle. B means pentagon. Edge means hexagon and S means square. So these are the letters for that shape. And from that we get to know the shape of insert.

1-Shape of Insert

Regular shapes		Irregular shapes	
T	▲	C	(80° 2 Corner)
R	●	D	(55° 2 Corner)
L	■	V	(35° 2 Corner)
P	⬟	W	(80° 3 Corner)
H	⬣		
S	■		

And in irregular shapes we have. See if we modify square insert by giving 80 degrees to its two corners. Then that insert it C shape insert what's in D. So if we give 55 degree angle to that two corners, then that shape is denoted as d. In v we have that two corners of 35 degree. If you have seen we named you insert which is much longer. So two corners of that insert are of 35 degrees and in w hexagonal shape insert with three corners of 80 degrees. That means it is modification of hexagon. So from our first parameter that means from these alphabets we get to know shape of insert. Just by looking at these alphabets. So for example watch t and mg insert. So T means triangular shape. That means insert will have triangular shape. Our second parameter is insert clearance angle. So if we draw a straight line from tip of the insert. And if we measure that angle from straight line to the face of that insert, then that angle is a relief angle means insert clearance angle. So here smaller angle will have larger surface contact. And larger surface contact will help to remove more material. As large surface contact will create large pressure. So as you can see n is having zero degree clearance angle. And due to this zero degree clearance angle, this insert is able to do total surface contact with material. And due to this large pressure will get applied and this will help to remove more material. If we see z it have 30 degree clearance angle. So in this case only smaller portion of insert will come in contact with material and do it which it will not able to apply more pressure. And that's why it will not able to remove more material. So here in symbol insert that means zero degree clearance angle insert is used or prefer for roughing process. As you know in roughing we have to remove more material. So there zero degree insert is preferred. And for finishing some clearance angle insert is preferred. As due to this clearance angle we get better surface finish.

2-Insert Clearance Angle

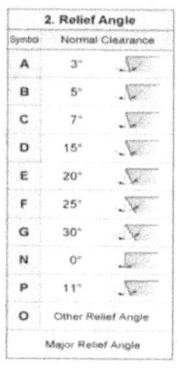

- Roughing process requires maximum pressure.
- So to achieve this insert must have maximum area of contact.
- So 'N' is preferred in roughing process.
- More finish requirement, more will be clearance angle.

So for roughing we use n that is zero degree clearance angle and for finishing we prefer some clearance angle. But remember more clearance angle will remove less material. So use this clearance angle insert according to your surface finish requirement. So this is all about insert clearance angle. Our next parameter is tolerance grade. As you can see we do not get exact size after machining for all the jobs. We will get some plus or minus size. Then required. So how much plus or minus is allowed for that job is known as tolerance. So as you can see in diagram for diameter one we have plus -0.03 tolerance. That means 1.03 is also allowed and 0.97 is also allowed. That means any diameter from 0.97 to 1.03 is accepted here. So 1.03 is called higher limit and 0.97 its called lower limit. So this tolerance is for this diagram. So I will repeat once again. So when we do machining for any job some plus sides or minus size is given to that job as we do not get exact size all the time. And that is why some plus sides or minus sides is provided for the job. So that plus sides or minus sides is called as tolerance. Here we will learn one small concept. So as you can see one meter is equal to 1000 m m and one m is equal to 1000 micron. So these two are the conversions that you have to remember. So coming to the topic tolerance grade. So we have two grades M grade and U grade. In M grade we have plus - 0.05. That means plus 50 micron is allowed and -50 micron is allowed. And this is mostly preferred for turning. And you grade have plus -0.08. That means plus 80 micron is allowed and -80 micron is also allowed here. And this U grade is mostly preferred in milling.

3-Tolerance grade

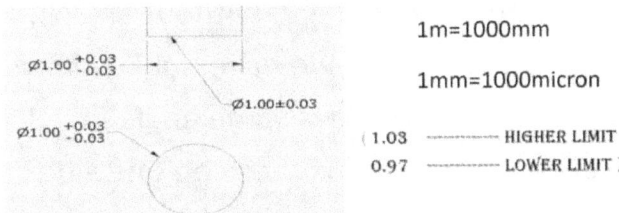

1m=1000mm

1mm=1000micron

(1.03 ---------- HIGHER LIMIT
0.97 ---------- LOWER LIMIT)

M-Grade ± 0.05 mm ------------------ Mostly preferred in turning
U-Grade ± 0.08 mm ------------------ Mostly preferred in milling

So we have these two tolerance grades. Our fourth parameter is Jeep breaker and clamping system. If you see insert carefully it have grew in starting or in corner. So this grow is there to break chip produce. That means during machining some material removed from job so that remove material is known as chip. And to break that material into small pieces this group is given. If we do not break that chip, then what happens? That chip will wrap the whole tool and do it to these. Insert will not able to cut material and we will get poor surface finish. And that is why chip is always break into small pieces. And to break that chip into small pieces we have glue on insert. Here we have so many types. So in this n, d and g are mainly used. If you understand these three types only, then also you can understand all these types very easily. So first we have n. So in n insert we will not have any hole. And chip breaker is also not therefore insert. So this is how it looks like. Simple insert without any hole and without any chip breaker. Our next type is D. So in T we have hole which is countersink hole. That means there is some angle in beginning of hole and chip breaker is there on one side only. And our third type is G. Here we have hole rigid cylindrical hole and g breakers are there on both the sides. That is double sided G breaker. So these are our chip breakers. Now you can understand any type given in that chart.

4-Chip Breaker And Clamping

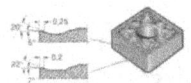 Groove for chip breaking.

Symbol	Hole	Hole Configuration	Chip Breaker	Figure	Symbol	Hole	Hole Configuration	Chip Breaker	Figure
W	With Hole	Cylindrical Hole	No		A	With Hole	Cylindrical Hole	No	
T	With Hole	One Countersink (40–60°)	One Sided		M	With Hole	Cylindrical Hole	Single Sided	
Q	With Hole	Cylindrical Hole	No		G	With Hole	Cylindrical Hole	Double Sided	
U	With Hole	Double Countersink (40–60°)	Double Sided		N	Without Hole	–	No	
B	With Hole	Cylindrical Hole	No		R	Without Hole	–	Single Sided	
H	With Hole	One Countersink (70–90°)	One Sided		F	Without Hole	–	Double Sided	
C	With Hole	Cylindrical Hole	No		X	–	–	–	Special Design
J	With Hole	Double Countersink (70–90°)	Double Sided						

-Mainly 'N', 'T', and 'G' is used.

So with the help of this you can know about insert chip breaker. And is there any hole in Insert or not. So you can know all these things with only symbols. So moving on. Our fit parameter is size of insert. So we call length of insert add size of insert sixth parameter are its thickness. So insert manufacturer do all testing and they decide insert size its thickness and all. So these values may vary. Insert to insert. Our seventh parameter is nodes radius. If you see insert carefully it is not sharp at the corners. It is somewhat rounded there. And that rounded surface is known as nodes radius. So it is mainly 0.2 or 0.4 or 0.8 or 1.2. Here. If insert have less nodes radius, then there will be less surface contact and this gives good surface finish. And if insert have more nodes radius then there will be more surface contact and it to more surface contact. It will remove more material. And that is why this is good for roughing. So less nodes radius insert are used for finishing, while more nodes radius inserts are used for roughing. So 0.2 or 0.4 M nodes radius inserts are used for finishing while 0.8 or 1.2 M nodes radius inserts are used for roughing.

7-Nose Radius

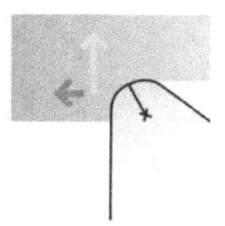

0.2/0.4 => (Finish)

0.8/1.2 => (Roughing)

Less nose radius => Less surface contact => Good finish

More nose radius => More surface contact => Good for Roughing

So these are our seven parameters. So let us see one example so that you can understand all these parameters very well. So our insert is re b m t 16 0404. So reads shape of insert. So if you remember we is 35 degree two corner. Insert that modification of square insert B its clearance angle. So if you look in that clearance angle chart B is having five degree clearance angle. After that M tolerance grade. So we have seen two tolerance grades M and u. So M is having plus -0.05 tolerance. And which is used for turning. And you have plus -80 micron tolerance. So we have already seen this. Our next parameter is D which is chip breaker. So we have seen that g breaker chart in which D is having one sided g breaker and countersink hole. So as you can see we are getting all detail about insert just by looking these parameters. So our next parameter is 16 rigid size of insert. That means length of that insert. So insert have 16 m length. Next is zero for rigid thickness. That means for m thickness and last zero for eight nodes radius. So insert have 0.4 m nodes radius. So let's find out whether these insert will be used for roughing or for finishing. So as you can see here we have clearance angle. That means it is preferred for finishing and zero for m nodes radius. It's also indicating us that it is preferred for finishing. Sometimes there is no surface finish requirement for job. So company you this same insert for roughing as well as for finishing. So whether to use that insert for roughing or for finishing it is totally depends on our requirement. But standard is that insert width clearance angle and 0.2 or 0.4 nodes radius are preferred for finishing process. Now you have to do one thing here. I have given some examples for you. So try to find out all details about these inserts. If you get any difficulty then you can prefer charts which we have seen. So try with this. So just by looking these parameters we are getting all characteristics of that insert. And from these we will get to know that whether that insert is preferred for roughing or for finishing. So this is all about insert. Now we will see external tool holder selection. So we have taken and insert in insert selection here also we will take one external tool holder b w l n r 2525 M06. So here we have nine parameters. That means for selecting extra inner tool holder we have nine parameters. So first is insert clamping. Second is insert shape. Third is approach angle. Fourth is clearance angle fifth is cutting direction sixth is shank height, seventh is shank width, it is two length. And ninth is cutting edge length. So let's see all these parameters in detail. So our first parameter is clamping method. That means how external tool holder will hold the insert. And according to that they have created some methods. So in this D it clamp with center hole. That means insert will have hole in center. And tool holder will clamp that insert with the help of that hole. Like this. And it's double clamp. That means clamping insert with two sides. As you can see here, clamping is from two sides. Next is s screw clamp. That means insert clamp with the help of screw only.

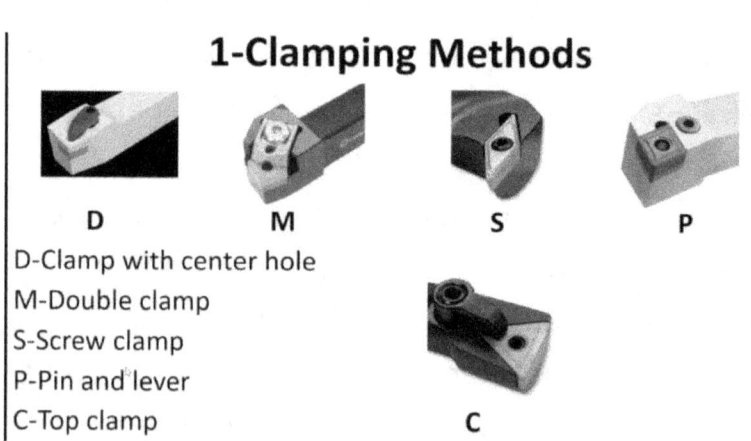

1-Clamping Methods

D
M
S
P

D-Clamp with center hole
M-Double clamp
S-Screw clamp
P-Pin and lever
C-Top clamp

C

Next is p pin and lever. So with the help of pin and lever we clamp insert in tool holder. And last is C top clamp. It's similar like clamp with center hole. But here clamping is not done with the help of center hole of insert. So these are our clamping methods. Our second parameter is insert shape that we have already covered. Our third parameter is approach angle. So these are approach angle. And we have clearance angle in insert. We have approach angle in tool holder. So in which direction tool is going to cut material. It is shown with the help of approach angle. So here we have so many approach angles in which and and J is mostly used. So remember their angle J is having 93 degree and L have 95 degree approach angle. Our next parameter is clearance angle that we have already seen. So moving on. Our next parameter is cutting direction. That means is it right hander tool holder neutral tool holder or left handed tool holder. So if we take this tool holder exactly in front of our eyes and if it is turning to right, then its right handed tool holder. If it's straight then its neutral tool holder. And if it is turning to left then its left hand tool holder. And after that we have shank height shank width and tool length. Here tool holders body is called as shank. So that body's height is shank height. Shank width is width of tool holder. And at parameter its tool length. So this total length is tool length here. Then divide it to length into grades. And this is standard. That means etch will have 100 m two length. If you want 2:07 p.m. length tool holder then you will not get this tool holder in market. You will get either 250 M or 300 m length tool holder. So this length is standard for particular grade. Our ninth parameter is cutting edge. As you remember in insert we call this length as sides of insert. But in tool holder it is cutting edge. So these are our nine parameters. So let's see one example. So here it is s v j p 2525 M 16 here. First parameter is clamping method. So as we will have screw clamp v is insert shape. So we is having 35 degree two corner insert. That means 35 degree two corner insert will fit in that external tool holder. After that we have j approach angle. So I told you to remember two types of approach angles. And and j l is having 95 degree approach angle. And J is having 93 degree approach angle be its clearance angle. So B is having five degree clearance angle. After that are these its cutting direction of tool holder. So if we hold tool holder in front of our eyes and if its turning to right, then its right handed tool holder or right handed cutting direction after that 25 which is 25 M shank height, next 25 is 25 M shank width and M is tool length. So M is having 150 M two length and last 16 is 16 M cutting length. Just like our insert sides we have insert length. So these are all parameters of external tool holder. So practice these I have given two more example for your practice. So try to solve them. After this we have internal tool holder selection. That means tools which are used for IED turning are internal tools. So to select that tools we have nine parameters. First is tool holder type. Second is shank diameter. Third is tool length. Fourth is insert clamping. Fifth is insert shape. Sixth is approach angle. Seventh is clearance angle. It is cutting direction and ninth is cutting edge length. So as you can see here, only first three are unknown remaining all we have already seen. So let us see these three in detail. So first parameter is tool holder type. We have two types of tool holder. First is steel with coolant hole. And second is solid steel. So steel with coolant hole. That means we have hole for coolant. As you can see in picture this type of hole are there in steel with coolant hole as this tool is used for ID turning. So during I returning large amount of heat get produce. And this may damage our tool. So to avoid this we have that coolant holes. And next type is solid steel. It don't have that coolant holes. Our next parameter it shank diameter. Now you know what is shank. So diameter of that shank is our second parameter. Our next parameter is tool length. So this is tool length. And this is also divided into some grades. So look this chart for tool length and all other parameters we have already covered earlier. So let us see one example for ID tool holder a 32 SBWINR06. So A is steel bar with hole for coolant. That means hole is present in this tool holder. Our next parameter is shank diameter. So diameter of that shank is 32 m. Third parameter is tool length. So s means 250 m two length p is pin and lever clamping system. So we have seen this in external tool holder w its insert shape. So its 80 degree three corner insert and its approach angle. So I have 95 degree approach angle. We have G as well which is having 93 degree approach angle and its insert clearance angle. So n is having zero degree clearance angle. That means it will use for roughing R its right hand cutting direction or right hand tool holder. And 0686M cutting length. So except for first three remaining, all we have already learned. Now try with these two examples. So from now, just by their names, you will get to know every detail property of that tool or that insert. So here we have completed nomenclature of that insert as well as tool holder. Now you will say this is only nomenclature of tool holder or insert. That means we are getting every detail about insert or tool holder just by looking at their names. But how to select tool or insert. That means if we have had material at that time which insert, great, we have to choose. Or if we have soft material at that time which insert grid we have to use. So how we can decide this. So you get this knowledge like which insert great is useful for hard material or regions that is useful for soft material. So you get all this knowledge only by experience. If you see there are so many insert grades available in market and to study, each insert grade is not possible. If you see each company have decided their insert grades, that means which inside is going to use for hard material and which insert grade is going to use for soft material that they have already decided. And they use all these insert grades according to that. So you have to node which insert grade your company is using for hard material or for soft material or different kinds of materials. And this is the only way to get knowledge about insert grades. If you go to study each insert grade then you will require much longer time. So I am telling you the shortcut way to get all these knowledge of insert grades. So now you will ask how company decides about these insert grades. That which insert grade will be used for hard material or different kinds of materials. So see company who manufactures inserts. They know everything about their insert. So assessment of that company visits different companies and they new information about their inserts. That means whether their insight is good for hard material or for soft material and all these information and also they offer some free inserts to company for trial. And company is getting free inserts. They accept it. And if company gets good results from that insert, then they continue to use that insert or that insert grade for that material. So this is how insert grades are already decided by companies for different materials. So you only have to collect all these information about insert grades. And you will get this knowledge only by the experience. So from now you have to see which insert grade your company is using for which material. So here we have learned nomenclature of insert and tool holder as well. And I told you one shortcut way to get knowledge about insert grades. So this is all about nomenclature and selection of insert grades and tool holders. Our next topic is insert failures and its remedies. As you are using insert, it is definitely going to wear out after some jobs. But sometimes insert gets wear out so early. So at that time you have to check some parameters. So here I am going to show you some insert failures. And I will also give you some solutions to avoid these failures. So our first failure is flank rear. This is how flank rear looks like. You may have seen this type of rear. If you get this we are after many jobs. Then its okay as these insert have some particular life. But if you get these rear so early then require job then there is some problem with your machine or insert.

Insert Failures and its remedies

1.Flank Wear

Cause
•High cutting speed
•Low wear resistance carbide grade

Remedy
•Use harder cutting tool material
•Reduce cutting speed
•Increase coolant pressure and check coolant direction

So what are these cards. So that we are getting these. We are so early. So see high cutting speed material cutting speed maybe high then preferred or lower your resistance carbide grade. That means your insert may have low grade which is not suitable for that job or material. So these are because for this type of we are. Now let us see its remedies so that we can avoid these type of rear. So use harder material grade for that insert. So that insert will easily cut that material off job. Also reduce cutting speed. That means we give cutting speed with the help of feed. So reduce feed to get better life of that insert. And third remedy is to increase coolant pressure and check coolant direction. So you have to check whether coolant is properly directed on your workpiece or not. If coolant is not properly directed on workpiece, then lots of heat will get generated and they may cause this wire. So properly direct that coolant on workpiece to avoid this rear and also increase coolant pressure. Our next we and eighth grader. We are so it looks like this. So as you can see we are getting this. We are on a per side of insert. So this rear occurs due to excessive cutting temperature and pressure on the top of the insert. So to avoid this excessive pressure on a per side of the insert, we have to reduce cutting speed. Use more. We are resistance cutting tool material. That means use high grade insert. These two remedies we have already learned and increase coolant pressure and check coolant direction. This is also we have covered. So as you can see these three four common remedies are there to avoid. We are of insert. So whenever we are occurs you have to try with these common remedies. Our next project is plastic deformation and it looks like this. And this we are occurs due to high temperature. So what to do to lower this high temperature radius cutting speed. Check coolant direction. Increase coolant pressure. Here we have new remedy to reduce cutting depth. That means you have to reduce depth of cut. To avoid this we have. So these are common remedies which we have to use while operating machine. Next we are its edge fracture. So this whole corner gets damaged in this project. And this where it's caused due to excessive load on insert. So to avoid this we'll use high grade insert reduce cutting speed reduce depth of cut. Here they have suggested to increase corner radius. So these nodes radius we have already learned. So try to increase nodes radius. To avoid this we are and check tool stability. If vibration occurs. If any vibration present in your CNC then try to remove that vibration. Here I have taken some basic ideas that normally occurs in insert. So if you learn these four wires, then also you get an idea about how to deal with any wheel and how to increase insert slice. So here we have learned insert we are observing. So here we have completed our sixth topic.

MEASURING INSTRUMENTS

We used some instruments so that instruments we are going to learn here. So you are going to have a great use of these instruments during operating. So we are going to see two measuring instruments. First is vernier caliper. And second is micrometer. You may have seen vernier caliper in your company. So what is the use of vernier caliper. So it is used for measuring height diameter and depth. So you can measure all three with the help of vernier caliper. So let's see parts of vernier caliper. First part is main scale in which you can see scale on both side. Upper side scale is in inches and down side scale. It's in m m. Then we have vernier scale. In vernier scale also we have scale on a per side as well as in lower side. So as you can see here 50 line of vernier scale is matching with 49th line of mains scale. So from this you get at least count of vernier caliper rigid 0.02 M. So many times in interview they ask what is the least count of vernier caliper? Least count is nothing but minimum size that can be measured by instrument. So what is the minimum size that vernier caliper can measure. So vernier caliper can measure minimum 0.02 m. So 0.02 m is the least count of vernier caliper. And if you go for interview in large scale company at that time, they can ask you the formula to calculate least count of vernier scale. So you can get this formula very easily if you Google it. So try to remember that formula as that formula may ask in your interview. But mostly in interview they ask what is the least count of vernier caliper. So you have to say least count of vernier caliper is 0.0 2MM. After that we have locking screw. So it is used to lock the measurement. If you measure any size or job and that reading you want to show to other person at that time, you have to lock the reading with the help of locking screw so that your reading will not be disturbed. So do measurement and lock the reading with the help of locking screw. After this we have fine adjustment screw. This is fine adjustment screw to do fine adjustments. So that you can get very accurate size. In that case we use fine adjustment screw. As you can see last count of vernier caliper h 0.02 m. So that mostly it is not used for accuracy measurement. And that is why we rarely use this fine adjustment screw. After this we have depth measurement blade which is used for depth measurement. After this we have fixed dots and sliding dots. That means this jaw is fixed and these slides. And due to this we get measurement.

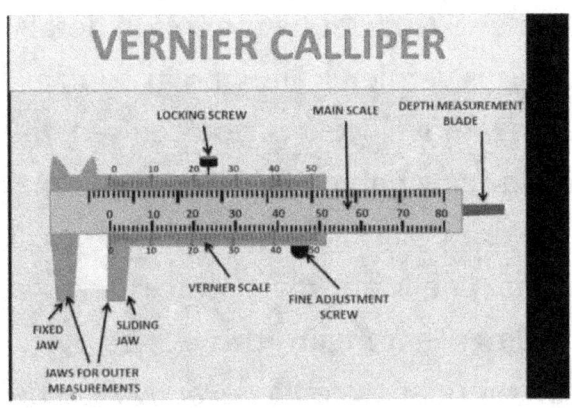

So these two jaws are used for outer diameter measurement. So all the outer measurements are done with these dots. And these two jobs are used for internal measurement. And with the help of the judge we do it. Measurement. So as I told you lower scale of main scale is in M and upper scale is in inches. So these all are the parts of vernier caliper. Now let us see how to calculate measurement with the help of vernier caliper. So for that we have one formula diameter to be measure. That means dimension we want is equal to main scale rating reading of main scale plus in bracket vernier scale reading into least count as this is in bracket. That means we have to do this calculation first. So reading of VS into. Please count what is least count of vernier caliper. It is 0.02 m. So this is the formula with the help of which we calculate measurement or reading. So see as soon as I hold a job in judge our moving jaw move and do it to which we are getting measurement. Now let us see this in detail. This is main scale and this is vernier scale. And as soon as I hold the job. See our jaw move. Now as you can see, this zero. So reading of main scale behind this zero is our main scale reading. So as we draw a straight line above zero. So reading behind that line is our main scale reading and vernier scale reading. That means these lines. Now you have to see which vernier scale line exactly matches with main scale line. So here 40 number line of vernier scale matches with line of main scale. So as we know our main scale reading, if you measure lines of main scale behind zero, then you will get 13 lines. So our main scale reading is 13 plus what we are scale reading. That means this 40 and least count which is 0.02. So put all these values in our formula. So our main scale reading is 13. Plus vernier scale reading is 40. And least count is 0.02. Now we have to do all these calculations. And whatever value we get that value is our measurement of that job.

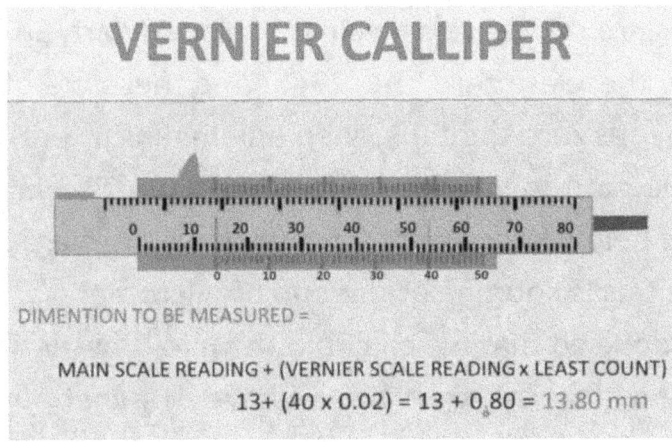

So we get 13.80 M. So this is size of that workpiece. So remember this formula and practice it so that you can easily measure any workpiece with the help of vernier caliper. So our next measuring instrument is micrometer. So micrometer is used to measure outer diameter of the job or workpiece. Now let us see its parts. So this is frame and on which we have anvil on both side. Second anvil is on spindle. And these anvils are so hard. That means these ends are hardened ends. And do it to which it does not. Well, as these two ends are always in contact with the material during measurement. So to avoid we are doing two continuous contact. These ends are made so hard and strong. And after that we have sleeve. This is sleeve. And on that sleeve we have scale for measurement. For measurement we have scale on sleeve. And scale is also there on this round part. So that is how these scales are on sleeve as well as on round part. So as you can see these are zero one, 2345. And these are of 0.5. That means zero then 0.5, then one means after 0.5 we have one. And so on means after 11.5 then 22.53. That means below lines are of 0.5 and above lines are of one m, and this is knurling at the end of micrometer. Knurling is done so that we can hold the micrometer properly. And after this knurling, we have a ratchet nut to check job. Precisely. That means if your job is touch with the anvil, then we have to use this ratchet nut to get accurate measurement. So let's zoom it to see this measurement. As we know these are 012345. These are upper scale. And in lower we have 0.5. That means 00.5. And after that what. And these are round scale. Return up to 50. And as I told you lower scale reading are of 0.5 M and above scale are of 1MM.

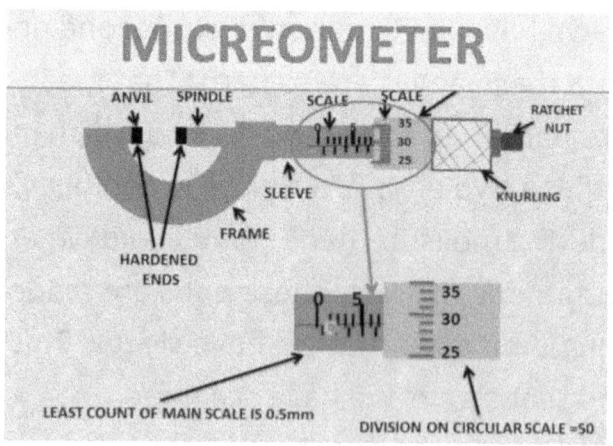

So this all the information about micrometer. Now let's measure one workpiece. So this is workpiece. So let's zoom its reading. So this is measurement. Now you have to see a per scale how much reading is visible on a per scale. So this five and six that means after five we have one line. So it's six. So this total is six M and is lower scale is visible. That means after six is there any lower scale line. So yes we have lower scale line as well. And as we know its value is 0.5 m. So total measurement of our micrometer will be this six plus 0.5. So total 6.5. Now we have to see this round skin. So which line of round scale matching with horizontal line. So it's 30 number line on round scale. Now let's put all these values in formula. So our formula is main scale reading plus thimble scale reading into list count. So main scale reading as we have calculated six plus 0.5. So it's 6.5. And thimble scale reading means this round scale reading which is 30 and least count of micrometer is 0.01. So after calculation we get measurement at 6.8 m. So size of that workpiece is 6.8 m. So that's how you can calculate size of any workpiece with the help of micrometer. So you will able to calculate all these measurement only when you practice. So here we have seen two measuring instruments vernier caliper and micrometer. So here we have completed our measuring instrument topic.

CAREER GUIDANCE 2

I hope you are understanding everything very well and I hope you work really hard for three months in CNC operating and then only you have watch for the projects. So now you are a CNC operator and you have watch two topics of setting. So now you have to become a CNC setter from CNC operator. And this journey will be of three months. So in this three months you have to cover setting topic as well as tool nomenclature topic. So you have to watch these two topics again and again. And whatever knowledge we have given here that you have to apply on your CNC machine as this is our setting topic. So you will require three months for it as here you are going to learn setting and this setting should become habit for you.

1. WATCH VIDEOS AGAIN AND AGAIN.
2. DO AS MUCH SETTING OF JOBS AS YOU CAN.
3. COLLECT INFORMATION OF TOOLS AND INSERTS.

And that is why you have to do so much hard work for next three months. The more you do setting, the more practice you will get. And along with that, you have to collect information about insert grades. So all this work you have to do in next three months means watch these projects again and again. Then do as much setting as you can and you have to collect information about tools and insert grades. So you have to do all these work for next three months. And in setting, if you are able to do setting then that is not over there. You have to add perfection in your setting. The more you do setting, the more perfect you will become and more accuracy you will get in your job sites. And that is why you need much practice in setting. And when you become perfect in your setting, then that is not over there. You have to improve your speed as well. So you have to do this both together. Firstly, try to add perfection in your work and then try to improve speed. So in this way you have to do so much work in next three months. So for next three months, watch these projects again and again and do as much setting as you can. Then bring perfection in it and then work on speed. And along with this collect insert great information. So you have to work on this for next three months. Our first to seven topic means operating plus setting. So you have to make these topics perfect. Don't rush. Try to bring perfection up to here. So you will require this much time. And if you are slow then you may require one more month. And that's okay. One more month will not affect that much. Only you have to focus on perfection and speed.

G-CODES

So from here our programing syllabus starts. So our eighth topic is G code. So what is g code. So it is preparatory function used to define operational condition. That means it's ABCd of programing language. So with the help of G course we can write a program. And we can also instruct CNC machine that where to move how to move and all. So all these we can do with the help of G codes. So let's see G codes. So these all are our G codes. If you Google G course then you will get so many zip codes and they may confuse you. That is why I have created list of G codes. And with the help of these G course you can create or develop any program. So here I have listed all G codes. G00, G01G02. So all codes are listed here. So let's learn all the G codes in detail. So our first code is G00. So whenever tool is not cutting any material at that time we have to move that tool fastly. So as you can see here from P1 to P2 two it is not cutting any material. So here during P1 to P2 we can move tool fastly. So here use G00 code. So remember if tool is not cutting any material at that time we have to move tool fastly. So let's see G00 code with the help of project. So here in this project I have used G00 code. And I am only trying to show you G00 movement. You only have to see whether tool is moving fastly or not. With the help of G00 code. So it's moving fastly. Our next code is G01. So whenever tool is cutting material at that time, we have to use G01 code. So during actual cutting we have to move tool slowly with the help of G01 code. And how slow it have to move is given with the help of feed. That means what will be the cutting speed of tool. It's given with the help of feed. So let us see G01 code with the help of project. So here I have used both G00 and G01 codes so that you can get difference between these two codes that G00 move fastly and G01 code moves slowly.

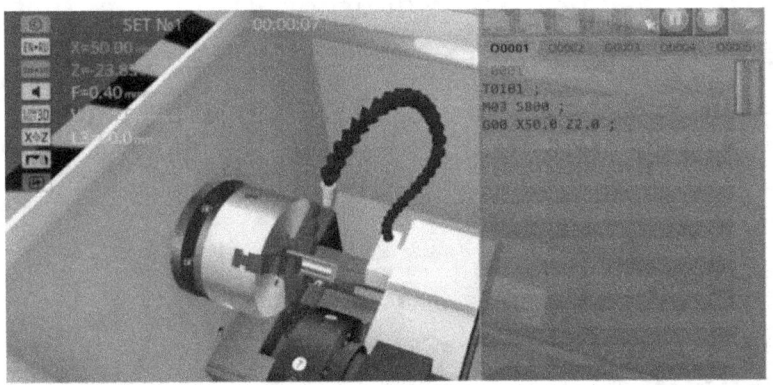

You can see it one second. And if you can see I have given feed 0.4 after G01 code. So it's compulsory to give feed after G01 code. If you don't give feed after G01, then tool will move fastly and do it. To which there are chances of accidents. So always remember you feed after G01 code. So here you have seen the difference between G00 and G01 code. So next code is G02 code. So if you want to cut the material in clockwise direction at that time use g zero to code. And if you want to cut material in anticlockwise direction then use G03 code. Here we are going to see two conditions. If tool is going to cut the material from a per side and second condition is from bottom side. Normally to cut some material from upper side but mostly in small C and C tools are arranged in front of each other. On this same bench. So at that time you have to consider tools direction. That means whether tool is coming from upper side or from bottom side. So if tool is coming from upper side, then the first radius that tool is going to cut is in anticlockwise direction. That means that first radius is in anticlockwise direction for tool which is coming from upper side. So use G03 which is for anticlockwise direction and second radius is in clockwise direction. So there we used g zero to code.

G02 CODE AND G03 CODE

- These code is used to define radius direction.
- G02 is for Clockwise direction. ↻
- G03 is for Anticlockwise direction. ↺

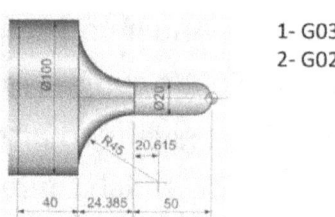

1- G03
2- G02

So this is for that tool which is coming from upper side. And if tool is coming from down side at that time first radius that it will cut it in clockwise direction. And for clockwise we use g zero to code. And second radius is in anticlockwise direction. And for anticlockwise we use G03 code. So you have to look this very carefully that from where your tool is coming to cut the material, where the tool is coming from upper side or from down side. So now let's see this with the help of project. So here I'm using downside tool. And here I have used G02 code. So G02 means clockwise. So C. Tool cuts the material in clockwise direction. So now let's see G03. That means anti-clockwise cutting direction. So I have written G03 code in program. So C tool cuts the material in anticlockwise direction. So I hope you understand the difference G02 is clockwise cutting direction and G03 it anti-clockwise cutting direction. So before using G02 and G03. Look at the tool from where it is coming. And according to that you have to use G02 and G03 code.

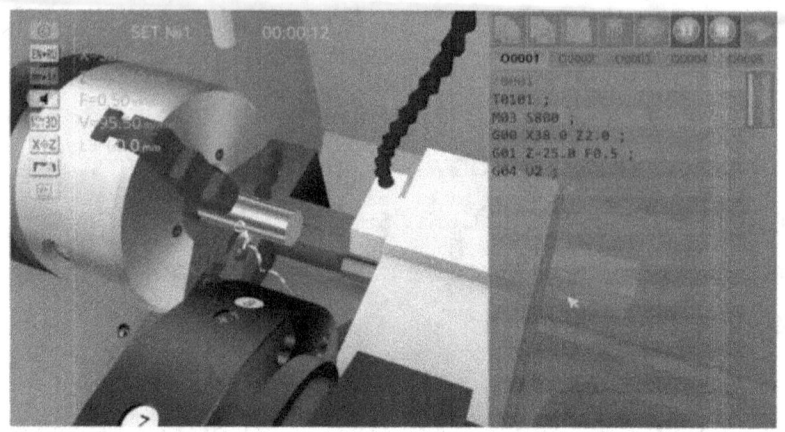

Our next code is G04. It is dwell time. So if you have to stop the tool for some seconds, then you have to use this code. So for better surface finish we stop tool for some seconds. So you have to write you after G zero for code. And then give how many seconds you want to hold that tool. So let's we want to stop tool for two seconds. So write G04 and then write you two. So machine will understand that it have to stop for two seconds. So let us see this with the help of project. So here I have written G zero for code to stop machine for two seconds. So see machine stops for two seconds and again starts running. So this is mostly use in drilling and grooving for better surface finish. Our next code is G 20 and G 21. G 20 code is used for size in inches and G 21 code is used for size in M. That means if you want all your dimensions in inches, then use G 20 code. And if you want all your dimensions in M, then use G 21 code. Here I have written one conversion formula. One inch is equal to 25 point 4MM. So you have to know this conversion formula. Our next code is G 28, G 29 and G 30 code. So as you know G 28 code is used for tool reference. That is for homing. But sometimes we have job with small operation like only chamfer or radius. So at that time, to take that tool to its home position and again call it to perform only that small chamfer or radius. So this will take much time. So at that time create a new home position near job and make it at reference position. And call that tool from that new reference position to perform that small operation. So this will save your time. So G 29 and G 30 I use to create new home position or new reference position for tool. So tool will stop there instead of going to its main reference position. So let us see this with the help of project so that you can understand much better. So firstly we will say G 28 reference position. So see this eight reference position for G 28. Now remember this position. Now we will see G 29 and G 30 reference position. So I have created new reference positions here. So this is the reference position for G 29. And this is the reference position for G 30. So here we have created two new reference positions. So with the help of these we can reduce our cycle time. So if your job is having only 1 or 2 operations like chamfer already is at that time, you can use G 29 or G 30 to create new reference position. But it is preferred to use G 28 for reference tool. Our next code is G 40, G 41 and G 42. So these are related to insert nodes radius effects.

G40, G41 AND G42 CODE

- To avoid tool nose radius effect on job size.

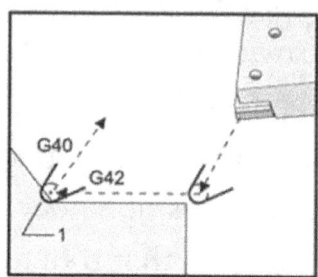

We will see these insert nodes radius effect in detail in our separate topic. So moving on. Our next code is G 50. So we have to tell our machine that what is the maximum spindle speed of our CNC machine. And about that spindle speed our spindle will not have to move. So to tell that speed we use G 50 code. Here I have written G 50 speed 2000. That means maximum speed of spindle is 2000 rpm. And about this speed spindle we will not able to rotate. So here we have created limit for spindle speed. With the help of G 50 code. So it is important to use this code in our program. If you don't use this code spindle rotate with very high speed. And this may cause accident. So it is better to use this code as it will reduce machine vibrations as well as it will increase to life. So always use this code in our program. I will tell you how to use this code in program in our standard format topic. But for now just remember that G 50 code is limit of maximum speed of spindle. Our next code is G 90 absolute mode. Now we will learn how to calculate coordinates with the help of this diagram. And here we will calculate coordinates with the help of absolute mode. So in absolute mode we do calculations from job zero point. Here our job 0.80. So by considering O as reference point we will calculate all the quadrants. So this method is absolute method. So in this course we are going to use this method. So let us see this method and let us find coordinates. So our first point is O. So O is our job zero point. As we have created job zero point during setting. So our coordinates of O are zero zero. Now we will do calculations from this point. So A what is x value for a. That means what is the diameter at point A 28. So x is diameter. And as you can see at a we have 28 diameter. And what about Z. So there is no movement in Z. So our coordinates for point A are 28 and zero. That means x distance is 28 and z distance is zero from all point. Now we will see b. So we will calculate coordinates at b from O as its absolute mode. So what x distance we have moved from O. So as we know x is diameter. So 28 diameter B have moved from O. And how much z distance B move from O. So it's -70 as is. Move in left side in Z. So it's minus. So our b coordinates are 28 and -70 x is 28. And z is -70. Now we will move towards c. So diameter in C is 32 and z distance from O is -70. So coordinates of C are 32 and -70. Now let us see these coordinates. So x in D that means diameter at d is 32. And how for our d from o in z direction. So it's 70 plus 80 which is 150. So -150 M. So coordinates of D are 32 and -150 x is 32. And z is -150. So this is how we have to calculate coordinates with the help of absolute mode. So I hope you understand this absolute mode. So now let us see increment mode. And for this we use G 91 code. And in this mode preceding point it's considered wide. Calculating coordinates. So here coordinates of O will be same. This is zero zero. What about A. So A have 28 diameter from O. So x at 828. And how far a move from O in Z direction. So A have not moved any distance in Z from O. So z coordinate is zero. So our code in case of A are 28 and zero. Now let us see b. So how far we travel in X from a. So b have not traveled any distance in x from A. So x coordinate of b is zero. As b and a are on same diameter. So in increment mode we consider preceding point naught. Oh okay. And how much z distance b travel from A. So it's -70. So coordinates of B are zero and -70. Now let us see. See. So how far we move in x from B. That means how much diameter c move from b. So its total for diameter and c do not move any distance in Z from b. So z is zero. So coordinates of C are four and zero. Now let us see d. So how much distance d move in x from c. As you can see they have not move any distance in diameter from c. So x coordinate of D80. And how far d move from c in z direction. So in z d move -80 from c. So z coordinates are -80. So coordinates of d are zero and -80. So this is our increment mode. So you can use any one mode from absolute or increment to calculate coordinates. Here in this course I'm going to use absolute mode as it easy to understand. And it is with very less confusion than increment method. But you can use any one from these two methods. So our next code is G 95. It is pretty unit. So with the help of G 95 we will tell CNC machine that our feed is in m m per revolution. That means in one revolution how much distance tool have to move is given with the help of feed. And for this we use G 95 code. So this G 95 code is mainly used in turning. That is in CNC turning. We just read in M per revolution. Our next code is G 96 and G 97. So as we know, as diameter changes, spindle speed also changes. So if you want all these speed calculations to be done by a CNC itself, then you have to use this G 96 core. So if diameter of job is changing then you have to use G 96 code along with cutting speed. How to calculate this cutting speed is we are going to see. But for now just remember you have to give G 96 code along with cutting speed. So if you want all the speed calculations to be done by CNC itself. So you have to use G 96 code along with cutting speed. So that CNC machine will understand that diameter is changing in workpiece. And according to that CNC will do its calculation. So this is G 96 code. You had to write G 96 code along with cutting speed. And if workpiece is not having any diameter change at that time you have to use G 97 code. So with the help of G 97 code, machine will understand that there is no diameter change in workpiece. So machine will not change its spindle speed. So you have to write G 97 code along with spindle speed that you want to give to your spindle. If you grew 1400 then spindle will rotate with 1500 speed. So G 97 is for fixed rpm, but in G 96 spindle speed will change according to diameter of the job. And in G 97 you have to give speed which is fixed for that workpiece. So these are our g course. So with the help of this g course you can make program of any job. So you have to remember these g codes. So here we have completed our G codes.

M-CODES

So our next topic is M codes. So what is important. So M code is machine code. That means this codes are related to machine parts. That means if we want to own the coolant then we use M08 code. If we want to rotate spindle then we use M03 code. That means these codes are related to machine parts. And you will get these codes written on your CNC machine as well. So here I have listed M codes M008 program stop. To stop program we use M00. Code M01. Code is optional. Stop this code we have already seen in control panel. There we turn on optional stop key along with M01 code and 038 clock white spindle rotation. If we want to rotate spindle clockwise then we use M03 code for anti clockwise spindle rotation. Use m zero for code M05 is used to stop spindle and zero six is used to change the tool. We use this code in our old

CNC machine, but in new CNC machine there is no need to write this code to change the tool. We can directly change the tool in new CNC machine. N07 is for high pressure coolant on to use coolant with high pressure. We use M07 code and M08 is used for normal pressure coolant. It does not have that much pressure. And zero nine is for coolant of M ten for shock. Clamp. If there is any need to clamp jug automatically by machine, then use M ten code and m 11 for check Unclamp m 910 eight spindle rotation of this code we have use in jar removal process. At that time we will log the spindle and then remove jobs. So at that time we use m 19 code M 38. End of program.

List of M CODE

M CODE	FUNCTION
M00	Program stop
M01	Optional stop
M03	Clockwise spindle rotation
M04	Anticlockwise spindle rotation
M05	Stop spindle rotation
M06	Tool change command
M07	High pressure coolant on
M08	Coolant on
M09	Coolant off
M10	Chuck clamp
M11	Chuck unclamp
M19	Spindle orientation off
M30	End of program
M98	Sub program call
M99	Sub program cancel

So program ends when you give M that D code. That means the LC machine will understand that program is ended here. After looking at M 30 code M 98, its sub program, called Small programs that will repeat in your main program are written separately. And to add that small programs in your main program we use M 98 code. And to cancel that small program we use M 99 code. This M 98 and 99 code are not used that much in CNC, turning this mostly use in VMC machine because in VMC some process repeats continuously. At that time. We use that small part of program separately and then add that small part of program in main program with the help of M 98 code. But in CSC machine we have cycles for that repeating program. So we rarely use these m 98 and M 99 code in CNC. Turning. So let's learn these m course with the help of project. So first m code is M00 which is program stop. So here I have written M00 code. So let us see whether machine stops or not with M00 code. See? It's working. So M00 is for program. Stop. Our next code is M01, which is optional. Stop. So after particular operation, we have to use this M01 code. So by looking at that M01 code machine stops if optional it's on so severe weather machine stops or not with the help of M01 code. See machine stop. So it's working. So M01 stops machine after particular operation. Our next M codes are M03, M04 and M05. These codes are related to spindle rotation. So M04 is for clockwise rotation. M03 is for anticlockwise rotation. And M05 is to stop spindle rotation. So it's working. That means our M codes are correct. Our next code is M062 will change. So before tool change, we have to write this M06 code. So it's working. Our next code is M07, M08 and M09. These calls are related to coolant. So M07. It's high pressure coolant. M08 is normal pressure coolant. And M098 is coolant of. Our next code is M 19 code. As we have already seen this in our jar removal process. So see this once again how spindle locks with M 19 code. So our all M codes are working properly. So these are the M codes. You will get these M codes on your CNC machine as well. So don't take that much tension of these M codes. So here we have completed our imports.

STANDARD FORMAT

We have seen G codes and M codes. Now let's see how to use these codes to make program. So to write this program we have one standard format. And with that standard format you have to write program. So that is standard format. We are going to learn here. So let's see standard format. So first we have to give name of our program. Here we are going to see program in Fanuc as well as in Siemens. So first we will see standard format for Fanuc control panel. So in Fanuc we have to give program file name. And here we have to start name with. Oh and after or write any for numbers. So this we have already seen and you can see it once again we have start with. Oh and then any four numbers after that we have N1 which is first operation. That means we are starting our first operation and after N1 we have to write G 28 U 0.0 W 0.0 means homing. Many times tool stops in middle of C and C after end of program. Or if someone forget to do homing at the end of program at that time. For safety purpose do tool homing in starting of program. After that we have to call tool with which we are going to perform operation. So in all CNC machine we have to write M06 and then give tool number. But in new C and C machine you can directly call tool just by giving its number T020 to first, zero to its tool number and next zero to its opposite number. So here we are writing two number tool to perform first operation. So that is why we have written T0202. So two number tool with offset in two number. And after calling tool we have to give some instructions to spindle. So we have to tell spindle that you can rotate maximum up to this speed. That means we have to give some speed limit to that spindle. And for that we use G50 along with spindle speed. How can we get the spindle speed that we will see later. But you have to remember that after tool call you have to give limit of speed. After that we have to give some cutting speed along with G 96 code. If diameter is changing in your workpiece, then we have to use G 96 code along with cutting speed. How to calculate this cutting speed. This we will see later. But you have to remember that you have to use D 96 code along with cutting speed. And after that M03 code to rotate spindle. So these are the codes of cutting speed and spindle direction. If we are job do not have any diameter change at that time. You have to replace G 50 and G 96 codes with G 97 code. And along with this code you have to write spindle speed and then M03 code. So you have to replace these two lines with this line when your required workpiece have constant diameter. So these are these spindle instructions. After that we have to give some rules to CNC machine. So first rule is G 90. That means we are telling machine that we are using absolute more here G 21 means we are taking all dimensions in M and G 95 means we are taking feed in M empire revolution. So this is our rule. This rule and we have to tell to our CNC machine. After that our main program starts. So firstly we have to call tool near our job from its reference position. And this moment will be fast movement. As during this movement there will not be any cutting of material. That is why we will use G00 code and call our tool near job. And after that we will turn on the coolant. So our tool is near our job and we have given all the instructions to the machine. So after this our actual program starts. So write all cutting program here according to your required workpiece. Just remember that if you are using G01 code then you have to give freely for it. So according to this feed tool will cut the material. So after completion of your material cutting program you have to leave the tool above your job surface. That means leave the tool in x and Z about your workpiece. And after that you have to stop spindle as well as coolant and then end of program. So that is how all the codes are written for one operation and for second operation we have to write here and two and then write all these codes with this format. So this is standard format to write any program. So you have to make any program with this format only. So practice this standard format as we are going to use this format for program writing. As you know we have written Cutting Speed after G 96 code.

```
O2222;                              Program file name
N1;                                 Operation Number
G28 U0.0 W0.0;                      Tool home position
T0202;                              Tool change command
G50 S___;                           Limit of speed
G96 S___M03;      G97 S___ M03;    Cutting speed and spindle direction
G90 G21 G95;                        Rule line
G00 X__ Z__;                        Tool call rapid
M08;                                Coolant on
G01 X__ Z__ F__;
G02 X__ Z__ R__;                    Material cutting program

G00 X__ Z__;                        Tool out of job
M05 M09;                            spindle and coolant stop
M30;                                End of program
```

So let's see from where we get this cutting speed. So if you have seen your insert box, so on that insert box cutting speed is given by. We see symbol. So we have different we see for different insert grades. Now let us consider we have k grade insert and cutting speed. For k grade insert it is two 25 meter per minute. And we will use this cutting speed. So we will write this cutting speed along with G 96 code. Now let us see how can we get maximum spindle speed that we have used along with G50 code. So to calculate that maxim spindle speed we have one formula. And this is the formula V is equal to pi d and divided by 1000. We each cutting speed by 3.14. And this pi is constant which is 3.14 d is diameter. We have to take minimum diameter of our required workpiece and its maximum spindle speed. That means speed that we use along with G50 code. So let's rearrange this formula and is equal to 1000 multiplied by. We divided by pi into d. So we have to find this maximum spindle speed. So we have to put all these values in our formula. So let us see how to do this with the help of one example. So here we have taken cutting speed at two 25 meter per minute. And these are our formula that we have recently seen. So let's put all the values in this formula. And it's equal to 1000 into cutting speed. So our cutting speed is two 25 meter per minute. So put this 225 value in formula divided by pi by value 3.14 which is constant. So put this value as well. Multiply by d d is the smallest diameter of workpiece. So here our smallest diameter is 28 m. So we put this value in formula. And after that we have to do this calculation. And after this calculation we get 2559 r.p.m.. So we will take approximately 2500 r.p.m.. So this is our maximum spindle speed. So we have calculated maximum spindle speed. And we also learn how we can get cutting speed. Now all the values are parameters from standard format are known to you. Now you can make program of any job.

So this is standard format. And we have seen this format in FANUC control. Now we will see standard format for Siemens control. So if you see the standard format is almost similar for both the control panels. Only little difference is there in both control panel. So here the difference is shown in red color as in final control program file name starts with oh and then any for numbers. But in Siemens we don't have any rule for giving program file name. You can give any name to your program starting with any alphabet or any number. Anything you can give here after that N1 first operation. Then in FANUC we use G 28 U 0.0 W 0.0 for two reference. But in Siemens we use G 74 or G 75. In some Siemens, G 74 is used, and in some Siemens G 75 is used. So you have to use this code according to your control panel to do tool referencing. After that we call D tool. So in Fanuc we write T0202 for tool calling. But in three months we write T02D2 for zero two is for tool number and d two is for offset number. Only difference is in value that we use zero two. And in three months we use d two. After that we write g 50 code to give maximum spindle speed in Fanuc. But in Siemens instead of that G 50 code we write and I'm s is equal to and after is equal to. We have to give that maximum spindle speed. So machine will understand that this is our maximum spindle speed limit. After that G 96 cutting speed and M03 which is same as our FANUC control. After that we give rules and same rules we have to give for Siemens as well. G90 is used for absolute mode. Here we use G 71 which is for sides in M m. In Fanuc. We use G 21 code for sides in M, m and in Siemens viewed G 71. And after that G 95 feed in M revolution. And after that we call our tool near our job. That means we rapidly call our tool near our job. After that coolant on. And after that we write our actual cutting program. So all these cutting program course as same in both control panel. Only difference is here that when we use G02 or G03 code for cutting material in radius at that time, we have to give radius width c r is equal to. But in Fanuc we give radius with r. That's the only difference remaining. All codes are same in both control panel. After that we have to take tool out of our job. Then spindle stop coolant off and end of program. So remaining all codes are same. So I will revise the difference once again. Firstly you can give any name to your program in Siemens control panel. For tool referencing wheels G 74 or G 75 code instead of G 28 code in value, we use G 28, and in Siemens we use G 74 or G 75. In Fanuc we use T0202, but in Siemens wheels T02, D two. In Fanuc we use G 50 and then maximum spindle speed. But in Siemens we have to use L I'm S is equal to to give maximum spindle speed. In Fanuc we use G 21 for sides in M. But in Siemens we use G 71 code after that for cutting material in radius. We use are in Fanuc but in Siemens we use c r is equal to two. Give that radius. So that is how we have to write program. So we have seen is standard format in board control panel. So you have to write programs according to this standard format only. So practice it. And remember this standard format as we are going to use the standard format while writing programs. So here we have completed our standard format topic.

NOSE RADIUS COMPENSATION

So our next topic is nodes radius compensation. As I had told you that we are going to see G40, G40 one and G40 two cores in separate topic. So this is that topic nodes radius compensation. So let's learn what is nodes radius compensation. As you can see here. And as you know insert corner is not sharp at the end. It is rounded at the corner. That means it has some nodes radius at the corner. And do it to that nodes radius. We face some problem. So as you can see here in this corner it is removing some extra material. Our dimension is same but due to that nodes radius it removes some extra material here. And that is why we do not get exact size that we want. So we can solve this problem with two methods. Either we do all calculations by our side. That method is manual method.

- Two methods to overcome with this problem:
  1. Manual method
  2. Automated Method

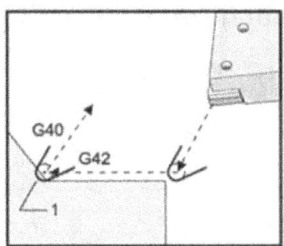

Manual Method:

Doing all the calculation by our side.

Automated Method:

All the calculations done by machine just by using G40, G41 and G42 code.

And second method is automated method, in which machine will do all these calculations with the help of these three codes? If we have to do all these calculations. Then what? What's the use of our CNC machine? And that's why we will tell machine to do all these calculations with the help of these three codes G 40, G 41 and G 42. So this effect that means is removes some extra material. Do it to nodes radius. So that effect is node's radius effect. And this effect we are going to see with the help of these three codes. So firstly we will learn G 42 code which is two nodes radius compensation two. Right. So if you hold this tool or insert exactly in front of your eyes and if that tool is working by moving in right direction, that means it is going to work in right side. Then we will use G 42 code. So see here our tool is going to cut material by moving in right side. That means it is going to work on already. So in this condition we use G 42 code. So for already use G 42 code for nose radius compensation. So if you use G 42 code then CNC machine will understand that tool is going to work in right side. And machine will do all the calculations by itself. Only you have to write insert nodes radius in geometry offset page.

G42 CODE(Tool Nose Radius Compensation To Right)

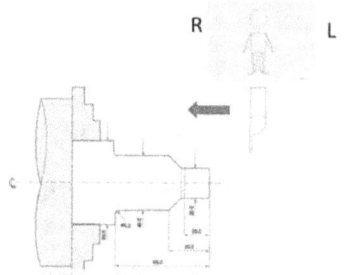

For OD, use G42 code for nose radius compensation.

It is compulsory to write insert nodes radius in geometry page as according to that nose radius machine is going to do all the calculations.

G41 CODE(Tool Nose Radius Compensation To Left)

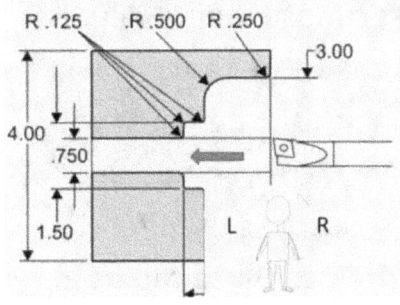

For ID, use G41 code for nose radius compensation.

That means if you write g what you do code, then machine will understand that it have to do nose radius compensation to write. So machine will find insert nodes radius for that tool. In geometry page and according to that nose radius machine will do all the calculations by itself. And you will get accurate size that you require. So you only have to remember that you have to give node radius in geometry offset page. So this is G 42 code two nodes radius compensation two write. Our next code is G 41. Code two nose radius compensation two left. So once again you have to take tool or insert exactly in front of your eyes. And then you have to see whether tool is going to cut material in left side or in right side. Here tool is going to work in IED means take tool in front of your eyes, and that tool is going to cut the material in left side. So in this case we are going to use G 41 code. That means for ID we use G 41 code for nose radius compensation. So as soon as you use this code machine will understand that it have to do calculation for two nose radius compensation to left. Then machine will see insert nose radius value in geometry page and according to that nose radius it will do all the calculations. And after using this code for ID you will get accurate size that you want. So if you notice G 42 code is used for returning and G 41 code is used for ID returning. And don't forget to write user nose radius in geometry page as machine. Do all the calculations with the help of that insert nose radius. If you don't give any nose radius in geometry page, then there will be zero zero in that nose radius column. So machine will do all the calculation by considering zero zero nose radius and you will not get accurate size that you want. So always remember to give insert nose radius in that geometry offset page. So these are our G 42 and G 41 code. And G 40 code is used to cancel these two effects. If your workpiece do not have any radius or chamfer at that time, there is no need to use this course as there will be no nose radius effect. Produce. This effect only produce when your job is having radius or chamfer. So if you use g 41 or g 42 for any tool, and you don't want this effect for another tool at that time, g 40 is used to cancel effect of G 41 and G 42. So here we have seen G 40, G 41 and G 42 codes. These codes are for two nodes. Radius compensation. Now we will see how to use this course in program. You have to know that how to use these codes in our program. So as you can see here in our rule line we have written G 40 code. That means all nodes radius effects before this code gets canceled. So we have to write g 40 in our rule line and in our actual cutting program we have to use g 40 or G 42 code. That means for overdetermined use G 42 code and for I returning use G 41 code. So you have to use any one code of these two codes. And then you have to write insert nodes radius in our geometry offset page. And after completion of material cutting program you have to use G for decoder. So that in effect of that G 41 or G 42 code will get canceled. So that is how we use G 40, G 41 and G 42 code in our program. If you go for manual calculation then you will require much time. But the main thing is that why we have to do these calculations. We have made a machine to reduce our time and do all the calculations by machine itself. So you G40, G 41 and G 42 codes for two nodes. Radius effects. So here we have given a note for you after writing G 40, G 41 or G 42 code, check whether you have given insert nodes radius in our geometry offset page or not. So always remember when you use g 40, G 41 or G 42 codes, then you insert node radius in geometry offset page. So this is all about nodes radius compensation. So here we have completed our nodes radius compensation topic.

FINDING CO-ORDINATES

Our next topic is finding coordinates. Until now, we have learned every concept of operating, setting and programing. So here we will learn our last concept which is how to find coordinates. It is very easy and basic concept, but as it is based off our programing. So we have added this topic in our course. So as you know x and z graph here I have shown you x and z axis in x. If we move upwards then it's plus. If we move downwards then it's minus in z. If we move right side then it's plus. And if we move left side then it's minus. So this is our x and z graph. And according to this graph we decide plus or minus. So look at this graph understand it and remember it. After that we will see how to find coordinates. And here we are going to use absolute method to find coordinates. So reference point. That means job zero point. So in setting we make this point as zero zero. After that we have a point. So at a point how much will be x. That means how much will be diameter at point A. So it's at T. And what about z. So there will be zero as a. Have not move any distance in Z. So coordinates of point eight are at t and zero. After that b point. So how much diameter is there at point B. So here we have at t diameter. And what about z. So B have moved 15 in z direction from job zero point. But B have moved to left side in z direction. So it's -15. So coordinates of B are at t and -15. Next point it's c so diameter at c it's 100. And see how move total 30 plus 15 which is 45. But as it is moving in left side so it's -45. So coordinates of point C are 100 and -45 x will have 100 and z will have -45 after that d point. So diameter and it's 100 and z distance is. If you add all these distance then it is 95. And as it is moving in left side so it's -95. So coordinates of point B are 100 and -95. So here we have calculated all the coordinates for paper job.

Find the co-ordinates

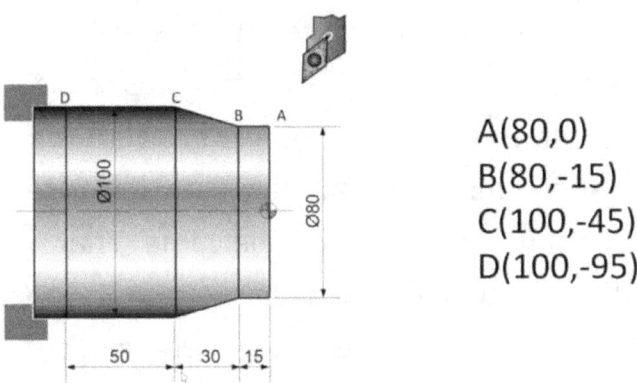

A(80,0)
B(80,-15)
C(100,-45)
D(100,-95)

This example was so easy as all the dimensions were already given. Now let's see one more simple example with radius. So see very simple example in which we have radius here also they had given all the dimensions. But I have two dimensions. So that you will use your brain to calculate these coordinates with the help of radius. Many times they will not give all the dimensions. So at that time you must be able to calculate all that coordinates. So at this point we have zero zero as its job zero point at point A x means diameter will be 30. And how much z distance a mu from. Oh, so it's zero. So coordinates of point A are 30 and zero means x is 30 and z coordinate is zero. Now be point. So how much will be diameter at point B. So this 30 plus radius of ten from a per side and radius of ten from down side. So that radius distance will become 20. That means this 30. And this total radius distance is 20. So total will become 50. So diameter at point B will be 50. And what about z. So B have mu distance equal to radius. That means ten. So B have moved ten. In z as our radius is ten. And as it is moving in left side so it's minus ten. So coordinates of point B are 50 and minus ten. That means x is having 50 and z is having minus ten. Now let us see see point. So at point C diameter will be same. That means 50. And how much distance we move in Z. So we have to calculate this distance here. So as you can see total distance is 100. And distance between point E and D is 50. That means distance between point D and O will become 50. And then only this total distance will become 100. So as we know, this distance between d and O is 50, and this distance between d to C is ten. And from this we will get distance between c to O. That will be 40. So coordinates of point C are 50 and -40. X is having 50 and z is having -40. Now let us see point D. So how much diameter is there at point D. So this radius distance becomes 20. So this 50 plus 20. So total 70. So diameter at point D will be 70. And what about Z. So e to d distance is 50. So this distance will also become 50. So coordinates of D are 70 and -50. Now let's see point E. So diameter will be same which is 70. And E have move hundred from job zero point.

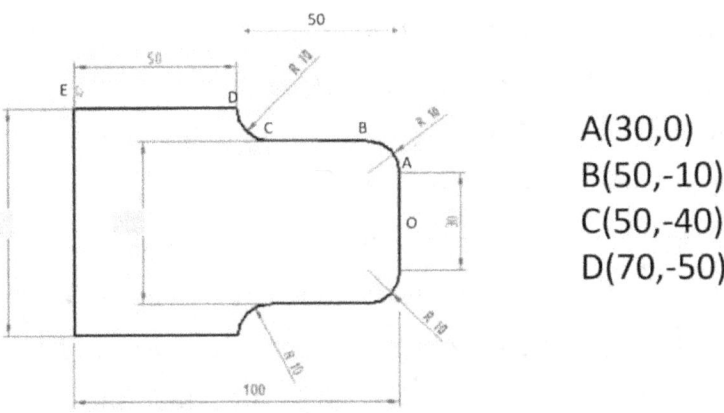

A(30,0)
B(50,-10)
C(50,-40)
D(70,-50)

So here we find out all the coordinates by doing some calculations. Now let's do some more calculations. As we have not done any calculation in town for example, as there all the values were known to us. But many times in chamfer angle is given instead of dimension. So at that time you must be able to find out coordinates by doing some angle calculations. So that angle calculations we are going to see here. So these are the three formulas for angle calculation sine cos. And then so sine is opposite side divided by hypotenuse. Opposite side means this angles opposite side divided by hypotenuse. So hypotenuse means this line and b means this slanted line hypotenuse and opposite side. Here it's x opposite 230 degree angle. So as you can see sine angle. So here we have 30 degree angle. So a sine 30 is equal to opposite side. So we have opposite side x divided by hypotenuse. We have hypotenuse value five. So we will write five. So sine 30 is equal to x divided by five. So if you see sine 30 value in calculator then you will get 0.5. So 0.5 is equal to x divided by five. And after solving this we will get x value as 2.5. That means this length will be 2.5. Similarly for cos angle cos is equal to this side divided by hypotenuse. So adjacent side means side which is adjacent to angle. So for cos cos 30 as we have 30 angle. So cos 30 is equal to adjacent side. So here our adjacent side is four means four is adjacent to 30 degree. So we will write four in place of adjacent side and hypotenuse. So we have hypotenuse value five. So we will write this five in formula as well. So our formula is cos 30 is equal to four divided by five. So this is cos angle. Here all values are known to us. Our next angle is then angle ten is equal to opposite side divided by adjacent side. Here opposite side is x and adjacent side is four. So 1030 is equal to x divided by four. If we see the value of tan 30 in our calculator then it's 0.577. So if we solve this then x will get 2.309. So here we have learned how to calculate this. X value means diameter value. And we learn all these three formula tan cos and sine.

Basic Calculation Formulas

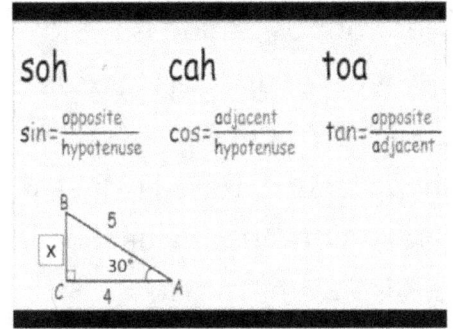

Sin (30°)=X/5
0.5=X/5
X=2.5

Cos(30°)=4/5

Tan(30°)=X/4
0.577=X/4
X=2.309

So with the help of these three formulas we have to do chamfer angle calculations to find diameter or length of this job. Now let us see one example. And in that example we will see how we have to do this chamfer angle calculations. So see this example. Here they have given this chamfer value add two into 45 degree. So let us find all these coordinates. Our job zero point which is zero zero that we all know after that a point. So how much diameter is there at point A. So it is six. And what about Z at A is not move any distance in Z. So z will be zero. So coordinates of point A are six and zero. Now be point. So as you can see there is no diameter given. And instead of that diameter they have given two into 45 degree. So with the help of this we have to find diameter here. This two is this length. So they have given this length at two. As you can see here this two is this length and 45 is this

45. Now we have to find this distance. This small distance we have to find. So to find this distance we have three formulas. First each sine. So a sine is equal to opposite side divided by hypotenuse. Here opposite side is x and hypotenuse. So we don't know hypotenuse here. So we can't use this formula. Our second formula is cos. So cos is equal to adjacent divided by hypotenuse. Here I just said which is two. But we don't have hypotenuse value. So we can't use this formula as here. Also we have two unknowns x and hypotenuse. So our third formula is ten. So if we use tan formula here so tan is equal to opposite divided by adjacent. Here opposite side is x and adjacent side it's two. So here we have only one unknown value. So we can find it. So let's put all these values in that formula. So it is 1045 is equal to x divided by two opposite side divided by adjacent side. So let's rearrange this formula. So we get x is equal to 1045 into two. And if you see this tan 45 value in calculator then it's one. So one multiply by two. Which is two. So we get this x distance at two. And we also get the same distance down side here. That means upper side two and lower side two. So total four and diameter of a is six. So if we add four in diameter of a then we will get diameter of b. And b have move two distance in z. And this two is this length. So z coordinate will be minus two. So coordinate of point b are ten and minus two. Now point C we have same diameter. At point C that means x will be ten. And c have moved seven distance from job zero. So z will be minus seven as it is

moving in left side. So coordinates of point C are ten and minus seven. Now point D. So diameter at point D is 20 as they have already given this value. And what about z. So if you calculate this distance five plus seven which will become 12. And as it is moving in left side. So eight minus so coordinates of point D are 20 and -12. So here we have calculated coordinates with the help of angle as well. So here we have learn how to find coordinates. So here we have completed all the concepts. So from next topic we are going to do programing for different examples. So we will see that examples in our next topic. So see you in next topic.

EXAMPLES (PART 1)

In this topic, you have to apply all the knowledge you have learned until now. So here we are going to see some examples. And with the help of these examples you will able to do programing. In this topic we will first see how to make program for finishing cut. And then we will see how to make program for job from raw material as in company. You will get a raw material and from that raw material you have to make program for your job. After that, we will see some operation like drilling, grooving, threading. And in last we will see one example in which we will use all the operations and we will make program for that job. And we will run that program in Fanuc as well as in Siemens control. So let us start. So first we will see how to make finishing cut program. So this is our diagram. So first we will find coordinates for this job so that we can make this program very easily. We will go slow as we are going to make program for first time. So coordinates at point A will be 16 and zero at point B. So if you do this calculation two into 45 that we have already seen how to do this calculation, I will give you hint for it. Here you have to use that angle for this calculation. So we will get coordinates at point B adds 20 and minus two. At point C our coordinates will be 20 and -22. This total distance is -22. Coordinates at point D will be 30 and -22. Any point coordinates will be 30 and -52. F point coordinates will be 50 and -72, and d point coordinates will be 50 and minus one zero two. Okay. So here we have calculated all the coordinates for this job. And now let's start to make program for it. So as we have already seen standard format to write a program. So we will use it here. So firstly we have to give program file name. Here we are going to make program in final control. So we will follow its rules. We will see next example in Siemens okay. So in file look program file name starts with. Oh and then for numbers as I have written here after that N1 first operation. After that we will call our tool T0101. After that we will do tool homing. For safety we reference our tool with G 28 U 0.0 W 0.0 code. After that we will give some instructions to our spindle means its maximum speed. So I will give here 2000 rpm G 50 speed 2000. After that we have G 96. As our job diameter is changing. That is why we are using G 96 code along with cutting speed 200 and M03 means here we are rotating spindle. After that we have run line G 90, G 21, G 95 and G 40. So this is rule line. After that we rapidly call our tool near our job. So for that we will write G00X 0.0 z 2.0. That means we rapidly call our tool here. After that we turn on the coolant. And for that we used N08 code. And if you notice we have written g 40 code in our rule line to cancel all the tool nodes radius effects. And now we are going to use G 42 code as we will do already turning here in right side. And that is why we are writing G 42 code here. Now you all know this G 40 G 41 and G 42 codes. And after that we will start our actual cutting program. So I will attach the tool to the job face. And for that I will write G 42 G01X 0.0, z 0.0 and feed 0.3. That means we are touching the tool to the job face. And as I have already told you, you have to use G01 code along with feed. Now we have to go at 0.8. And for that we will write G01X 16.0 as our diameter at point A is 16 and there is no need to write feed for every G01 quarter. If you don't give a feed here, then machine will consider previous feed and do material cutting with that previous feed. After that we have to go to point B and if you have calculated B points coordinate then that will be 20 and minus two. So we will right here G01X 20.0 z -2.0 okay. So our two reach at b point. Now we have to take our tool 2.3. And for that we will write G01Z -22.0. So we reach at point C. Now we will go from point C to point D. And for that we will write G01X 30.0. Then we will move our tool from point to point E. And for that we will write G01Z -52.0. Then we will go from point A to point F, and coordinates of point F are 50 and -72. So we will write G01X 50.0, z -72.0. And then we will take our tool to point G. And for that we will write G01Z minus one 02.0. So that is how we have to do cutting of material. So here we have completed our cutting. After that we will lift our tool above the job surface. And for that we will write G00X 70.0, z 2.0 and along with this we will write g 40 code to cancel node's radius effect. And after that we will do tool homing by writing g 28 yields 0.0 W 0.0. And after that we will write M05M09 and m 30.

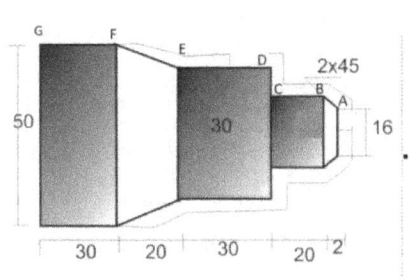 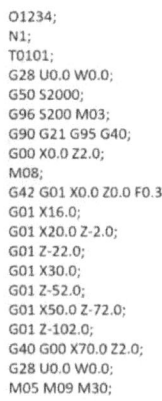 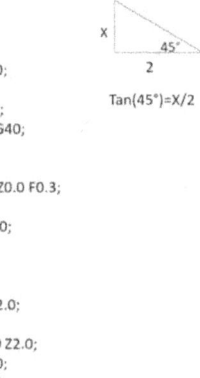

That means spindle stop, coolant off and end of program. So that is how we have to write finishing cut. So I hope you have understand how to write finishing cut program. So we have seen this finishing cut program in Fanuc control. Now we will see how to write program in Siemens Control. And we will take an example. Having radius in it. So see this is our workpiece. So let us start writing program for finishing cut. So firstly we will give program file name Bush one as we are working in Siemens. So we can give any name to our program. After that N1 first operation. After that G 74 or G 75 X 0.0 Z 0.0. This is for tool reference. And after that we will call our tool TSC 01D1. And after that we will instruct spindle. And for that we will write I I'm s is equal to 2000.

IN SIEMENS

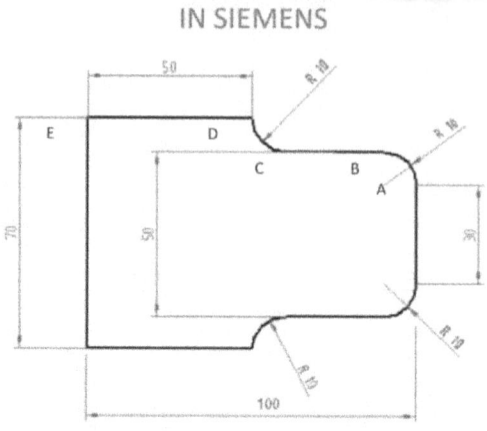

```
BUSH1;
N1;
G74/G75 X0.0 Z0.0;
T01D1;
LIMS=2000;
G96S150M03;
G90G71G95G40;
G00 X0.0 Z2.0;
M08;
G42 G01 X0.0 Z0.0 F0.3;
G01 X30.0;
G03 X50.0 Z-10.0 CR=10.0;
G01 Z-40.0;
G02 X70.0 Z-50.0 CR=10.0;
G01 Z-100.0;
G40 G00 X100.0 Z2.0;
```

Maximum spindle speed is 2000. After that G 96. Speed 150. This is cutting speed and M03. After that our rule line G90 G 71, G 95, G 40. So G 71 means we are taking sides in M m and after that we will rapidly call our tool. And for that we will write G00X 0.0, Z 2.0. So here we rapidly call our tool near our workpiece. After that coolant on. And for that we have M08 code. And after that our actual cutting program will start. So firstly write G 42 two nose radius compensation. After that G01. We will lightly touch our tool to did your face. So for that we will write G01X 0.0, z 0.0 and feed 0.3. After that we will go up to point eight, and for that we will write G01X 30.0. After that we will go at point B. And if you see this curve is anti-clockwise so we will use G03 code. And along with G03 code we will write B points coordinate which are 50 and minus ten. And along with this you have to give radius which is ten. And for that we will write c r is equal to 10.0. After that we will move to point C, and for that we will write G01Z -40.0. After that we will go to point D. And if you see this curve is clockwise and for clockwise we use g zero to code. So we write here g zero to x 70.0, z -50.0. And c r is equal to 10.0. As this curve is of ten radius. So we will write c r is equal to 10.0. And after that we have to go to point E. So we will write G01Z -100.0. After that we will lift our tool. And for that we will write G00X 100.0, Z 2.0. And along with this we will use G 40 code. So nose radius effect will be canceled. And after that we will do tool homing. And for that we will write G 75 x 0.0 Z 0.0. And after that we will write M05M09 and m 30 okay. So here we have seen this Radius program also. So we have seen two programs and made a finishing cut program for it in both Siemens as well as in FANUC control. But in company you won't get to make a program for finishing gate only. There they will give you raw material. And from that raw material you have to make program for your required job. So here we will take one raw material, and from that raw material we will make program for one diagram. So this is our raw material diameter 101 multiplied by length 85. This length is 85. And from this raw material we have to make this workpiece. And you can't make this job directly. So before this job you have to make these type of job. That means firstly you have to remove this extra material okay. So we will make this type of job first. So first we will find these coordinates. So if you see in diagram this radius is 16. And similarly this will be here also. So total 32 and this diameter is 35. So if we add this 35 and 32 then we will get our required points. So that addition is 67. So this point is 67 okay. Now let's start to make program for it. So first write program file name then operation number. After that we will do total homing. Then we will call our tool. Then we have to give instructions to this spindle with the help of G, T and G 96 codes. Then we will write our run line. And after that we will call our tool rapidly near our job. So if you see here I have rapidly call our tool at x 100.2 and Z 2.0. Now you will see why we have called our tool here. So if you see our raw material we have one M extra material and we will not make that 100 diameter in our first cut. Firstly we will do roughing and after that finishing and for finishing we are going to give zero point 2MM material here. And that is why we are calling our tool at x 100.2 and Z 2.0. And after that we will take a cut in z direction. So I will write G01Z -85.0. And along with this I will write g 42. And if you see this total distance is 82. If you add this distance then you will get 82. But I have taken 85 for safety purpose. So we will take a cut in z direction up to -85. And after that we will lift this tool. And for that we will write G00X 102.0 z 2.0. So we lifted our tool. And now we have to take our tool down. So we move our tool down in X direction. Here I have taken large cuts so that program length will be reduced. But you don't do this. If you are making program on machine then you have to take maximum one M or two m cuts. If you take large cuts then this will damage your tool as well as machine. So always take small cuts here and explaining concepts. And that is why I am taking large cuts. But you have to take small cuts only. So here I call tool at x 96.0. And now how much cut we have to take in z direction. So we will not take total cut in z as we want this diameter 100 here. So that's why we will take cut in Z up to -35.8. As we are going to give 0.2 M material here for finishing. And that is why we will take cut in Z up to -35.8. And after that we will lift that tool above job surface. And for that I have written here G00X 100.0, Z 2.0. After that we have to take that tool down up to x 92.0. And then again we will take it in Z up to -35.8. Then again we will lift that tool up to x 100 and z 2.0. Then again we will take that tool down up to x 88.0. Here I have taken large cuts, but you have to take small cuts only okay.

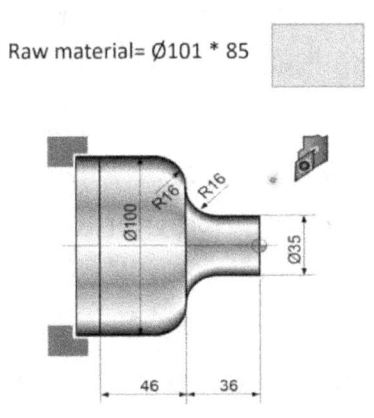

```
O1234;
N1;
G28 U0.0 W0.0;
T0101;
G50 S2000;
G96 S200 M03;
G90 G21 G95 G40;
G00 X100.2 Z2.0;
G42 G01 Z-85.0 F0.3;
G00 X102.0 Z2.0;
G00 X96.0;
G01 Z-35.8;
G00 X100.0 Z2.0;
G00 X92.0;
G01 Z-35.8;
G00 X100.0 Z2.0;
G00 X88.0;
```

Then again take cut in Z up to -35.8. Then again lift that tool up to x 100 and z 2.0. Again take down up to x 84.0. Then again take cut in z up to -35.8 and then again lift that tool up to x 100 and z 2.0. Then again take down. So you have to do this again and again. So here I will tell you last cut. So we will take that cut again and again. And we will reach up to x 67.0 as our this point is 67. If we take this tool below 67 then this radius portion will cut. And we don't want this. So we will take that tool up to x 67.0. And then we will take cut in Z up to -35.8. Then again lift that tool up to x 100 and z 2.0. Now we will not consider this radius part. We will make this radius part later. So if this radius length we minus from this 36 then we will get 20. That means this distance is 20. And after 20 we have this radius part. So up to 20 we will make our job like this. This type of job we are going to make here. And that radius part we will cut later. So our tool is at x 100.0 and z 2.0. So we will take that tool down up to 63.0. And then cut material in Z up to -20. If you take more cut in Z then that radius part will damage. So you have to take that cut in Z up to -20 only. And after that, lift that tool up to x 100 and z 2.0. And again you have to take that tool down. Then take a cut in Z up to -20.0. Then again lift that tool up to x 100 and z 2.0. Again down again take cut in Z again lift that tool. So this process will continue. So I will tell you last cut in this. So take tool down up to x 35.2. So here we are going to give 0.2 material for finishing. So we are at 35.2. After that we will take a cut in Z up to -20.0.

And after that Z00X 100.0 z 2.0. So our this diagram is ready. That means we have removed extra material. Now we are going to cut extra material present near our radius part. So we will take two three cuts here to remove extra material so that there will not be any extra load on our finishing tool or inside. That is why we are going to remove this extra material. So you can remove this extra material in many ways. So here I will tell you one simple way to cut this extra material so that you can easily cut this extra material. So firstly you have to decide some points. So here I have decided this point Z -19.8 because we have already taken cut up to Z -20.0. So for 50 we will take cut up to z -19.8. So we will fix this point Z -19.8. And in that line we will move our tool down. And we already know this point is 67. So firstly we will call our tool up to x 60.0 and z -19.8. And after that we will cut material in radius up to x 67.0. And as we have already taken cut in z up to -35.8. So we will take cut up to -35.5. So here we have fixed some points means this z -19.8 and this x 67.0 and z -35.5. So now you have to take tool down from this point and take radius cut. So let us see this. So firstly we will call our tool up to x 60.0 and z -19.8. After that we will write Z02. As this radius is clockwise and our x is 67. As we have already calculated this. So we will write Z00X 67.0, z -35.5, and our 16.0 as our radius is of 60 M, then lift that tool up to x 100.0 and z 2.0. After that we will take tool down, and for that we will write x 55.0 and z -19.8. After that we cut material in radius. So we will write

Z02X 67.0, z -35.5 and r 16.0. Again lift tool up to x 100.0 and z 2.0. Again, take tool down in x up to 50.0. Again, we will cut material in radius x 67.0, z -35.5 and 16.0. Again, lift the tool again take it down. So we will continue this process. So I will tell you that last good here we have already taken cut in X up to 35.2. So we will call our tool above 35.2 for our safety. So we will call our tool up to x 35.5. And for that we will write G00X 35.5 Z -19.8. So our tool is here. And now we will use G02 for this radius. And for that we will write G02X 67.0 z -35.5 and r 16.0. And after that we will lift our tool G00X 100.0, Z 2.0. So as you can see I have fixed two points here. After that I take tool down in Z -19.8 and cut material in radius by using G02. So this is the simplest method to cut this material. You only have to fix two points. And then by taking tool down, cut that extra material in radius by using G02 code. So practice for such programs. So this is our roughing process. Now we will do finishing. So we will write. And to after that we will do homing of our one number tool. Then we will call our finishing tool and then rapidly call that tool near our job. So we will write G00X 0.0 z 2.0. After that we will lightly touch our tool to the jaw face. And for that we will write G 42, G01, z 0.0 feed 0.3. Okay, after that G01X 35.0 so that we will reach here. After that we have to take it in Z up to -20.0. And for that we will write G01Z -20.0. After that we have to make this curve. And for that we will write G02X 67.0, z -36.0 and add 16.0. After that, here is some calculation. Here. After this curve we have to go a little straight. If you do calculation here this 67 plus this radius distance which is 32 will become 99. But our diameter is 100. So we have one more diameter at straight line. So for that we will take tool one upwards in x. And for that we will write G01X 68.0. And after that we have this curve. So for that we will use G03 as this is in anti-clockwise direction. And for that we will write G03X 100.0, z -52.0 and r 16.0 as this curve is of 16 radius. And after that we have to take it in Z. So we will write G01Z -82.0. And after that we will lift that tool above the job surface. And for that we will write G00X 105.0 z 2.0. After that we will do tool homing. And after that we will write M05, M09 and m 30 and off program. So here we have taken one raw material. And from that raw material we write program for this job. So you need much practice to write programs for such job more. You practice this more easily. You will able to write programs okay. So now we will see one group program. So this is its raw material 54 into 150. So 54 is the diameter and 150 is the length of our raw material. And from this we have to make this job. And diameter of this job is 50. So we have four m extra material. So firstly we will remove for m extra material. So first is program file name then n1 then tool referencing. After that tool call. Then we have to give instructions to this spindle. After that ru line. And after that we will rapidly call our tool near job. So firstly we have to cut for m extra material. For that we will take our roughing tool and call it at x

52.0, Z 2.0. As we will do two roughing passes and we will give 0.2 material for finishing. So we will write G00X 52.0, Z 2.0. And then we will take in z up to -150. As length of our workpiece is 150. After that we will lift that tool up to x 60.0 z 2.0. And then we will take that tool down up to x 50.2. As we are going to get 0.2 material for finishing. So we are at 50.2. And then we will take cut in Z up to -150. After that leave that tool above the job surface x 60.0, Z 2.0. And after that take that to down up to x 0.0. As we are going to do finishing here. So firstly we will attach the tool to the jaw face with feed 0.3. After that we will take that tool upwards up to x 50.0. So we will write Z01X 50.0. And after that we will take a cut in z up to -150. And for that we will write G01Z minus one 50.0. So this is all about roughing and finishing. Now we will do grooving process. So we will call our grooving tool. And after that we will rapidly call our tool. So we will call our tool above 50 diameter. So we call it at x 55.0 and z -20.0. As we have created this phase of our tool at zero zero, that means we touch this phase here. And then we have put z 0.0 in geometry offset page and that is why this phase of our tool is zero. And that is why we have to take this tool up to z -20.0 to make groove. And that is why we write G00X 55.0 Z -20.0. And now we will start our actual cutting program. So firstly we will take our tool up to x 50.0. And then we will slightly lift our tool up to x 51.0. Because we do not cut this material directly in one cut we take small cut in x, then slightly lift that tool up. Then again we will take a small cut. Then again lift that tool. So that is how we do grooving process. If we cut this material directly in one cut then this will damage our insert. And that is why we take small cuts and lift it and again take small cuts. So that is how we do grooving. So we measure up to x 50.0. Then we will lift our tool slightly up up to x 51.0. Now we will take a small cut in x. And for that we will write G01X 48.0. So we are at x 48.0. Now we will lift our tool. So we will write G00X 49.0. And again we will take a cut in x. So we will write G01X 46.0. And we will again lift tool one upwards up to x 47.0. Then again G01X 44.0. Again take cut and again lift that tool up to x 45.0. So we will continue to cut this material in this manner to cut then lift it again to cut again lift it. So we will continue to do this. So here I will tell you one last cut. So as we want to make grooving diameter. So we will write G01X 30.0. And now we will lift our tool. So we will lift directly outside of our job. So we will write G00X 55.0. So we lifted our tool above the job surface. And this same groove we will make here as well. So we will call that same tool above these groove. And for that we will write G00Z -40.0. And then we will start our cutting process G01X 50.0. We will touch our tool to the jar face and then we will lift that tool. Then again we will take small cut and then again lift that tool. Here I have taken large cuts. But you have to take small cuts only. So continue doing this. So here I will tell you one last cut as we want groove diameter as 30. So we will write G01X 30.0. And after that we will lift our tool out of the job. And then we will do tool humming. So as you can see our grooving tool has same thickness as our groove. But this will not be the same condition for all the time. Sometimes our grooving tool thickness may not be equal to groove that we want. So let us see how to do grooving. In that case. So this is that diagram here we want EDM groove and our grooving tool insert having three thickness. So let's write program for it. Firstly we will write program file name then n1 then tool humming. Then we call our grooving tool. Then we will instruct our spindle. And after that rule line then we will rapidly call our tool. As our diameter is five here. So we will call our tool at x 6.0. And as you can see I have called tool at z -3.0 because our this phase is zero. And that is why to take that tool here we have to write z -3.0. If we write z 0.0 then our tool will stand here. And we don't want this tool here. We want that tool here. And that's why we have to write Z -3.0. And after that we have to take it in X direction by taking tool up down up down. So we will write G01X 3.0ft 0.1. And then we will lift our tool up to x 4.0. Then again we will take it up to x 1.0. Then G00X 2.0. And in last G01X 0.0. And as I have told you, we can hold this tool here for some seconds by using G zero for chord for better surface finish. So we will hold it for one second by writing G zero for u one. And now we have to lift this tool above the job surface. So we will write G00X 6.0. After that we will move this tool left side. But how much we have to move this tool in left. So as we have read up to Z -3.0. So we will not add -3.0 more in z we will add less than -3.0. Because if we add minus three more then there are some chances that some material will remain here. And that is why we will move our tool by minus two more. That means total G00Z -5.0. Then again we will take it in x direction by taking tool up down up down. And then we will take our tool out of this job. And now we will move that tool in Z direction. And as we have already reached up to Z -5.0. So now we will again add minus two m. And for that we will write G00Z -7.0. And then again we will take third index by taking tool up down up down. So our one material is remain. So we will lift our tool and then add minus one m in Z. So we will write G00Z -8.0. And then again we will take it in x direction by taking tool up down up down. Then in last we will lift that tool above the job surface. And then we will do tool humming. So that is how we do grooving. When our tool thickness is not equal to groove thickness or groove width. So here we learn how to do grooving. So we will see our remaining operations in next part.

EXAMPLES (PART 2)

We will see threading problems. So here I have taken one problem in which I have added glue. That means whatever we have learned until now. That we will cover here means roughing, finishing and grooming. And we will see threading as well. So our raw material its diameter 60 into 60. That means this diameter is 60. And this length is also 60. And as you can see job have threading here grooming. And we will use roughing and finishing to remove extra material. So first we will see threading. Thread depth means depth of this thread. So how to calculate this depth. And for that we have one formula which is 0.61 into page. So what is this page. So distance between this point to this point is our page. And this page is given here. So see here 1.5. So this is the page of this thread. So our thread depth will become 0.61 into 1.5 which will become 0.915. So this is thread depth. Means thread will have this depth. And thread diameter means how much will be our thread diameter. Means this minimum diameter we call it as minor diameter. So how much will be our minor diameter. So to calculate that minor diameter we have one formula. So thread diameter means minor diameter is equal to major diameter minus two into depth.

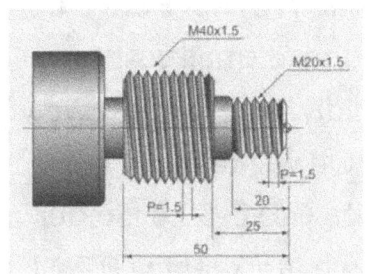

Raw material=Ø60 * 60

(Thread depth)=0.61*pitch
=0.61*1.5
=0.915mm

(Thread dia) = Major dia - (2*depth)
= 20 – (2*0.915)

So major diameter is maximum diameter. So here we have major diameter 20. So I will calculate minor diameter for this 20. So major diameter is 20. And depth which we have already calculated at 0.915. So put all these values here. And after calculate one we will get 18.17 M means this is minor diameter 18.17. So this is how you have to calculate this minor diameter. So let's write program for this job. And we will make this job from raw material. So firstly we have to make our raw material like this. We have already seen this in our last example. We make this type of job and then made radius and chamfer here. But in this example we do not have any chamfer or radius. So a simple example. So let's start it. So first is program file name then operation number. Firstly we will do roughing. Then we will do tool homing. After that we will call our roughing tool. Then we will instruct our spindle. And after that we have a rule line.

Then we will rapidly call our roughing tool near our job. And if we see this total diameter then it's 60. Here I am digging large cuts. But you have to take small cuts only. After that we will call our tool rapidly. And for that we will write G00X 55.0, Z 2.0. After that we will take cut in z direction. And for that we will write G01Z -59.8. So our this total distance is 60. But here we are keeping 0.2 M material for finishing. And that is why we are writing Z -59.8. After that we will take tool out of the job. So we will write G00X 70.0 Z 2.0. And then we will take cut again. And for that we will take tool down up to x 45. And then again we will take it in Z up to -59.8. Then again we will lift our tool x 70 and z 2.0. And then we will take our tool down up to x 40.2. As we are going to give 0.2 material here as well. And after that we are going to take cut in Z up to -59.8. Then again we will lift our tool up to x 70.0 and Z 2.0. That means our this part is ready. And now we are going to make this part. And for that we will take our tool down up to x 38.0. And after that we will take cut in Z up to -24.8. We will keep 0.2 M material here as well. So we will go up to -24.8. After that we will lift that tool up to x 50.0 and Z 2.0.

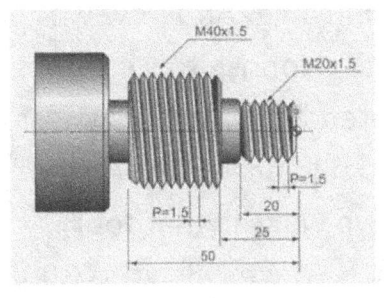
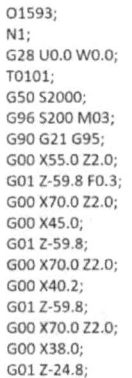
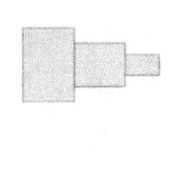

After that we will take our tool down up to x 34.8. And again we will take cut in Z up to -24.8. Then again lift tool x 50.0 Z 2.0. Then again take down. And then again take cut in z direction. Then again lift that tool. So this process keeps going. So here I will tell you the last cut. So we will take our tool down up to x 20.0. As we are going to keep 0.2 material for finishing. And after that we will take cut in Z up to -24.8. And then lift that tool up to x 50.0 and Z 2.0. And after that we will do the homing of that tool. So our this part is ready. And now we will do finishing. So end to end we will call our finishing tool. And now we will rapidly call our tool near job. After that we will lightly touch our tool to the jaw face. And for that we will write G01X 0.0, z 0.0 and feed 0.3. Then we will go up to x 20.0. And for that we will write G01X 20.0. After that G01Z -25.0. After that G01X 40.0. Then we will go G01Z -60.0. And after that we will lift our tool up to x 50.0 and z 2.0.

And then we will do tool homing. So up to this we have completed roughing as well as finishing. And now we have to make this groove. So let's suppose this group is of ten m. So we will take group instead of ten m thickness. And now we will call our tool. So n3 and T0303. After that we will rapidly call our tool. So we will call that tool slightly up in X and up to -60. In Z. So we will write G00X 70.0 z -60.0. So we are calling this tool slightly upside as there are chances of tool collision with this material. So we will call this tool up to x 70.0. And then we will start our actual grooving. So we call our tool at x 70.0 and Z -60.0. And now we will take a cut in x up to x 40.0. In this cut only small amount of material may remove as there is no material present here. After that we will lift our tool slightly up. So G00X 41.0. And then we will take a cut. Here I am taking large cuts. But you have to take small cuts only. So we take very small cuts in groove up to 0.5 only. But here I am trying to focus on concept only that how to make a groove. So once you understand this concept, then you can make this program with 0.5 MB cut. So here we will take cut up to x 36.0. And then we will lift that tool. After that we will take a cut again in X. And for that we will write G01X 32.0. And then G00X 33.0. We lift our tool. So you have to continue this process. Take a small cut then lift that tool again. Take small cut again. Lift it. So this continues. So here I will tell you one last cut. So we will consider this diameter as 20. So we will take a cut in x up to 20.0. And for that we will write

G01X 20.0. And then we will lift our tool up to x 70.0. And then we will do tool homing. So we have made this groove. Now we will make these two threads as our main focus. What to make these threads.

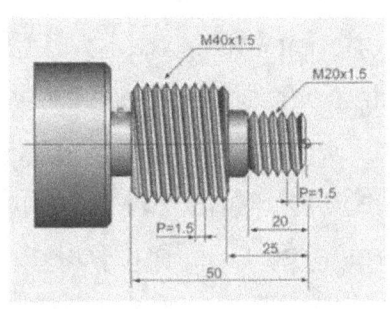

```
G01 X0.0 Z0.0 F0.3;
G01 X20.0;
G01 Z-25.0;
G01 X40.0;
G01 Z-60.0;
G00 X50.0 Z2.0;
G28 U0.0 W0.0;
N3;
T0303;
G00 X70.0 Z-60.0;
G01 X40.0 F0.1;
G00 X41.0;
G01 X36.0;
G00 X37.0;
G01 X32.0;
G00 X33.0;
G01 X28.0;
G00 X29.0;
G01 X24.0;
G00 X25.0;
```

So we have made our revision for roughing finishing and grooving. So let us see these threads. So firstly we will see this first thread. So N4 and then we will call our threading tool. Then we will rapidly call our tool. And for that we will write G00X 20.0, z 2.0. And after that if we take good by writing G01 code, then that tool will cut straight. Cut like our roughing or finishing as machine won't understand that it have to make a thread here. And that is why to make this thread we have G 92 code. So by using G 92 code machine, we will understand that it have to make thread here. And that is why we have to use G 92 code. And our x here is 20. And we want to make this cut inside head up to -20. And in threading feed is equal to pitch of that thread. That means whatever page they have given here we have to write that page in feed. And this is very important point. So page of the thread will come here in feed. So after that we only have to give cuts in x. Because this G 92 code will work in cycle. That means tool will take cut then lift up and then come outside of the job. So you don't have to write code for lifting tool and taking tool outside this process cycle do itself. So here we will write g 92 x 20.0, z -20.0 and feed 1.5. So tool will take cut lift up and then come out of the job. And now you have to give cuts in x. Here I am taking large cuts. But you guys take small cuts only. So I write x 19.6. So tool will come down and then follow G 92 cycle. That means it will take cut in Z then move up and come out of the job. After that we write x 19.2. So tool will move down. Then take cut in Z. Lift up and come out of the job. So this is this cycle. We don't have to write all the codes only give cuts in X. And that's it. Tool will cut material according to G 92 cycle. So here I will tell you one last cut. So here we have to take cut up to minor diameter. And we have already calculated minor diameter which is 18.17. So tool will come down up to x 18.17. And then tool will cut material. That means it will make threads according to G 92 code. And now we have to make thread here as well. So we will lift that tool. And for that we will write G00X 40.0 z 2.0. And after that G000 -24.8. So we stand our tool here. And after that we will write G 92 again as we want to make thread here as well. So we will write G 92 code. And then we will give dimensions of this thread exceeds 40.0 and z is -50.0. And feed which is equal to peach. And here peach is 1.5. So tool will take a cut in Z, then lift up, then come out of D job. Then we will give x 39.6. So tool will come down. Then it will take cut in Z. Again lift up and then again come out of the job. So that is how we have to write this cycle. So here what will be our minor diameter. So if we do calculation for this thread then we will get minor diameter at 38.17 m. So we will write x 38.17. So two will come down to 38.17. And then make this thread by following G 92 cycle. So here we made both of these threads. After that we will lift that tool out of the job. And for that we will write G00X 60.0 and z 2.0. And then we will do tool homing. So that is how we learn how to make thread. And we made this whole job from raw material. So remember these formulas as these are very important formula to find minor diameter. And without minor diameter you cannot make thread.

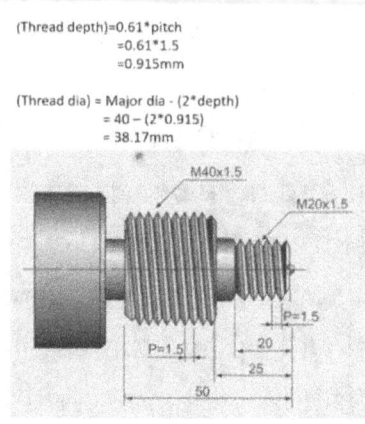

```
N4;
T0404;
G00 X20.0 Z2.0;
G92 X20.0 Z-20.0 F1.5;
X19.6;
X19.2;
X18.88;
X18.84;
X18.80;
......
X18.17;
G00 X40.0 Z2.0;
G00 Z-24.8;
G92 X40.0 Z-50.0 F1.5;
X39.6;
X39.2;
X38.8;
.........
X38.17;
G00 X60.0 Z2.0;
G28 U0.0 W0.0;
```

So remember it and practice it. After this we are going to see drilling. So here we will do simple drilling. So this is drill. And we have to make drilling up to Z -20.0. So let's see how to do drilling. So firstly call the tool G00X 0.0 and z 5.0. After that we will lightly touch our tool to the job face. So we will write G01Z 0.0 and feed 0.1. So always give let's feed for grooving as well as for drilling okay. So after that we will take a cut in Z up to -2.0 means as we have taken cuts in grooving means take a small cut, then leave that tool again. Take a small cut again, leave that tool. The same kind of cuts we have to take in drilling as well. So we have reached up to Z -2.0. Then we will move back. So we write G00Z -1.0. Then again we will take a cut up to z -4.0. Then again we will move back up to z -3.0. Then again we will take a cut up to z -6.0. Then again take it back up to z -5.0. So that is how we have to take these cuts. So here we have to make this drill up to Z -20.0. So we will write Z01Z -20.0. And after that we will take that tool out of the job Z00Z 2.0. And then we will do two roaming. So that's how we do drilling process similar like our grooving. The only difference is that we do grooving in X direction and drilling in Z direction. So until now we have seen all the operations and we learn these operations with example. So now we will see one example in which we will cover all the operations. So we will make program for that job. And we will run that program on Fanuc as well as on Siemens control. So this is that example. So look at the dimensions of this job or simply draw this drawing as you will require this drawing for making program. Here we have taken raw material as 34 into 60. So from this raw material we are going to make this job. So firstly we will do roughing and make our raw material like this. So firstly we will create this shape of our job. And this is our whole job as we will need these dimensions. So firstly we will write program file name. Then n1 means firstly we are going to do roughing. Then tool call then tool homing. Actually do tool homing first and then call your tool T0101. Then spindle instructions, then rule line and use G40 in it as chamfer is present in this job. So use G40. After that we will call our roughing tool rapidly as our raw material is diameter 34 into 60. So we are calling our tool up to x 32.0 and z 2.0. As we are going to take two M cut here. After that we will write M08 then G 42 compensation code. And then we will start our roughing. So we will write G01Z -60.2. We will take 0.2 more cut in Z for safety and then feed. So we have taken our first cut. Then we will lift our tool up to x 40.0 and Z 2.0. Then we will take tool down up to x 30.2. As we are going to give 0.2 material for finishing. So we will take tool down up to x 30.0. And then again we will take a cut in Z up to -60.2. Then again lift that tool and then again take down. So our this part is ready now. Only finishing it remain here. So now we don't take cut in Z up to -60. So if you check this distance then it is 50. And here also we will keep some material in Z. So we will take cut up to z -49.8. Then we will lift our tool x 40.0 Z 2.0. Then again take down up to x 26.0. And then again take cut in z up to -49.8. Again lift it x 40.0 z 2.0. So this process will continue. So here I will tell you one last cut. So we will take tool down up to x 20.0. And we are going to give some material for finishing. So if you can see this diameter it is 20. If you calculate this chamfer then you will get 20. So we will take tool down up to x 20.2.

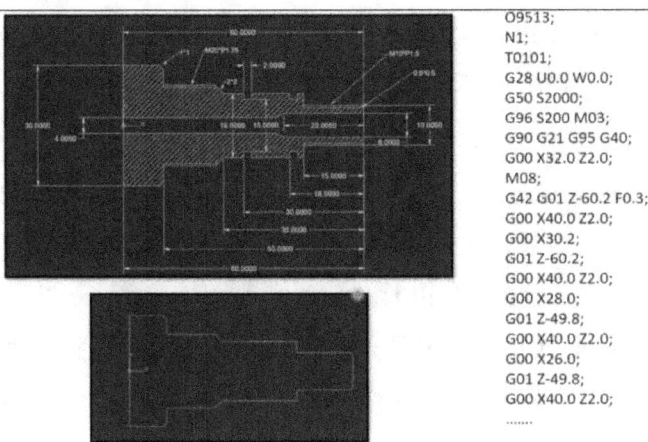

```
O9513;
N1;
T0101;
G28 U0.0 W0.0;
G50 S2000;
G96 S200 M03;
G90 G21 G95 G40;
G00 X32.0 Z2.0;
M08;
G42 G01 Z-60.2 F0.3;
G00 X40.0 Z2.0;
G00 X30.2;
G01 Z-60.2;
G00 X40.0 Z2.0;
G00 X28.0;
G01 Z-49.8;
G00 X40.0 Z2.0;
G00 X26.0;
G01 Z-49.8;
G00 X40.0 Z2.0;
........
```

As we are going to keep some material for finishing and then take a cut in Z up to -49.8. As we have to keep 0.2 material here as well. Then again we will lift our tool x 40.0, Z 2.0. Then again take tool down up to x 18.0. And now we do not have to take cut in Z up to -49.8. We will take cut up to this point only. And if you see this distance then you will get -35. So we will take cut in Z up to -34.8. As we will keep 0.2 material for finishing. So g0 one Z -34.8. Then again lift that tool up to x 40.0 Z 2.0. And then again take tool down. So this will continue until we reach near this diameter. So if you can see this diameter so it is 16. So we will keep 0.2 material. And that is why we will take tool down up to x 16.2. And then we will take in Z up to -34.8. Then again we will lift our tool. And then again take tool down up to x 14.0. Now we don't have to take cut up to -34.8. We will go up to here only. So if you can see this distance so it's -15. So we will take cut up to -

14.8. As we will keep some material here as well. So we will write D01Z -14.8. Then we will lift our tool x 40.0 z 2.0. Then again we will take tool down up to x 12.0. Then we will take in Z up to -14.8. Again lift it. So this process will continue until we reach near this diameter. And this diameter is ten. So we will take tool down up to x 10.2. And then we will take it in Z up to -14.8. And then we will lift our tool. So if you can see our this job is nearly ready. But we will have some extra material here as we have chamfer here. So we will remove this extra material so that there will not be any extra load on our finishing tool. So to remove this extra material, we will call our tool at x 16.5 and then -34.5. So as we had fixed some points in our radius example that similar way, we are going to fix some points here. So we fix one point in Z. So that point is z -34.5. As we have already taken cut up to z -34.8. So for safety we will fix point at z -34.5. And now we have to take good in z direction. So we will take cut Z01X 21.0 z -37.0. So we have taken a cut here means this step cut. So we take a small cut here so that this material will reduce. And now we will slightly lift our tool in X up to 28.0. And then we will move our tool in Z up to -49.5. Because we have already taken cut up to Z -49.8.

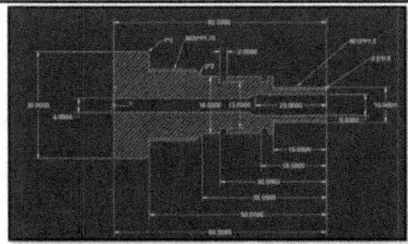

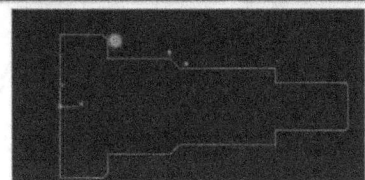

```
G01 Z-49.8;
G00 X40.0 Z2.0;
G00 X18.0;
G01 Z-34.8;
G00 X40.0 Z2.0;
G00 X16.2;
G01 Z-34.8;
G00 X40.0 Z2.0;
G00 X14.0;
G01 Z-14.8;
G00 X40.0 Z2.0;
G00 X12.0;
G01 Z-14.8;
G00 X40.0 Z2.0;
G00 X10.2;
G01 Z-14.8;
G00 X40.0 Z2.0;
G00 X16.5 Z-34.5;
G01 X21.0 Z-37.0;
G00 X28.0;
G00 Z-49.5;
```

So we stand our tool at Z -49.5. And now we will take a small cut here as well. And we will take cut slightly up in x as this diameter is 30. So we will take cut up to x 31.0. And z distance is -51. So we will take it up to -51. So our code will be Z01X 31.0 z -51.0. So we have taken a small cut here as well. So here we have removed this extra material as well. So now we will take that tool upwards in x up to 50.0. And along with this we will use G 40 code to cancel all compensation effects. And then we will do tool homing. So here we have completed roughing process. And now we will do finishing. So for the finishing and to then we will call our tool T0202. Then we will rapidly call our tool at x 0.0 z 2.0. And now we will slightly touch our tool to the jaw face. So we will write G01G 42 Z 0.0 feed 0.3. So we use compensation and we lightly touch our tool to the jaw face. After that we will write G01X 9.0. As you can see they have given that chamfer as 0.5 into 0.50.5 in top side and 0.5 in bottom side. So if you calculate this chamfer distance then it will be one m. So if you minus this one m from total diameter then you will get 9MM. And that is why we have taken cut up to x 9.0. And after that we will make this 0.5 into 0.5 chamfer. So for that we will write G01X 10.0 z -0.5. After that we will take cut in z. So we will write G01Z -15.0. After that we will write G01X 16.0. So this straight cut. Then we will move up to z -35.0. So we will write G01Z -35.0. After that we will make this chamfer. So for that we will write G01X 20.0 z -37.0. Because if you see this chamfer dimensions they have given two into two x distance is two. And z distance is also two. So top side two and bottom side two. That will be four. And this diameter is 16. So our total diameter will become 20. And that is why we have written G01X 20.0 z -37.0. That means we have added 2MZ distance in 35. After that we will take a straight cut in Z up to -50.0. After that we will write G01X 28.0. As you can see this chamfer one into one means top side one and bottom side one means total two. So if we minus these two from total diameter, then we will get x 28.0. And that is why we have written G01X 28.0. After that we will make this chamfer. So we will write G01X 30.0 z -51.0. So this diameter is 30. And this chamfer is of one m means z distance is one m. So 50 plus one is equal to 51. And after that we will take a straight cut. So for that we will write G01Z -60.2 0.2 is additional for safety purpose.

Then we will lift this finishing tool up to x 70.0 Z 2.0. And then we will do tool homing. So until now we have completed our roughing as well as finishing. So our this part is ready. And now we will make groove in this. So these two groove we are going to make. And that is why we will use two m thickness groove insert. And we have to call that tool rapidly about this groove. And this diameter is 16. That means about 16 m m. So m3 D0303. And now we will rapidly call our tool about 16 means up to x 16.5. And for z distance if you see here it is 18.5. So we will write z -18.5. So we will call that tool at x 16.5 and z -18.5. And now we know how to do this grooving. Here I will take large cuts. But you have to take small cuts only. So take small cut then lift that tool up again. Take small cut again. Lift it. So that's how you have to make this cut that you already know. So let us start it. So we will write G01X 16.0ft 0.1. And then we will lift that tool up up to x 17.0. Then again we will take good G01X 15.0. Then again we will lift that tool up. So that is how this cutting process will continue as our this diameter is 13. So our last cut will be up to x 13 D01X 13.0. And now if you want to hold this tool here for some seconds then you can hold it by using g zero for code. But I will directly lift it. So I will write G00X 16 0.5. Now this same groove I have to make here as well. And that is why I will move that grooving tool in Z above this group, which we are going to make. And Z distance for this groove is -30. So I will write G00, Z -30.0. And now we will make this groove by taking tool up down up down up to diameter 13. So write this code as you know this. Now and then lift that tool by writing G00X 17.0. And then do tool humming. So our the grooves are ready now. So now we will make threading here. So I have already taught you how to make thread. So before going for actual threading we will do some calculations. So here our first thread is M10 into pitch 1.5. So let us do some calculations for M10. And also we will find its minor diameter. And we will do the same calculation for 20 diameter as well. For 20 we have pitch 1.75. So we will do calculation for it. And for minor diameter. So you already know all these formulas. So you can easily find these minor diameters. So now we will start writing our threading program. So N4 we are going to make M10 thread. Then t0 4042 tool call. And after this we will rapidly call our tool. So as you can see we have called our tool at x 9.8 and z 2.0. That means we are going to take 0.2 M cut here. And that is why we have called tool at x 9.8. After that, write G 92 code thread cycle code by using this code, machine will understand that it have to do threading. And then we will give threading dimensions x 9.8, z -14.0 and thread 1.5 as our pitch is 1.5.

For dia=10mm

(Thread depth)=0.61*pitch
=0.61*1.5
=0.915mm

(Thread dia) = Major dia - (2*depth)
= 10 − (2*0.915)
= 8.17mm

For dia =20mm

(Thread depth)=0.61*pitch
=0.61*1.75
=1.0675mm

(Thread dia) = Major dia - (2*depth)
= 20 − (2*1.0675)
= 17.865mm

So since the machine automatically do threading here, that means that g 92 cycle means it cut in Z up to -14.0. Then it will lift up and come out of the job. And now we only have to give cuts. That means x. So we write x 9.5. So again it will take a cut lift up and come out of the job. Then again x 9.2. So this process will continue until we reach our miner diameter. So as we have calculated our miner diameter which is 8.17. So you have to write your last cut as 8.7 ten. So machine will run that code. And after that lift that tool. So for that I have written G0 zero x 19.8 z 2.0. Here we intentionally lifted this tool up to x 19.8. As we are going to do threading here as well. After that we will travel that tool in Z up to -35.0. Keep some safe distance. And after that we will start our threading process. So we will write g 92 x 19.8 z -50.0 as this length is 50 and feed 1.75 which is pitch of the thread. So tool will make thread up to z -50.0. Then tool will lift up and then come out of the thread. Then again we will give x 19.6. So tool will take up to z -50.0. Lift up and again come out of the thread. So this process will continue until we reach our miner diameter. And as we have already calculated our miner diameter which is 17.865. So we have to write x 17.865. So that tool will come at x 17.865. And then tool will make that thread then lift up. And then we will take that tool out of the job by writing G00X 25.0 Z 2.0. And then we will do tool homing. So here our threading is also completed. So now we will do drilling here. So for drilling we will write N5 then call our drilling tool. Firstly we are going to do this for Am drilling. So we will call for M drill. And then we will rapidly call that for m drill at Z 2.0. And after that you know how to do this drilling. Firstly take a cut. Then again move back. Then again take cut. Then again come back. So this you already know. So we will write G01Z 0.0ft 0.1. We slightly touch our tool to the job first. And then Z01Z -5.0. Here I have taken large cuts. But you have to take small cuts only. So we have taken cut up to Z -5.0. Then take that tool back by writing Z -4.0. Then again take cut z -10.0. Then again back z -9.0. So that is how I will take this total cut means this total length. And I will take a little extra cut in Z to remove all the material properly. So here we have total length 60 m m. So I will take extra cut. That means Z01Z -61.0. To remove all the material properly. And after that we will take that tool out of the job. So G0 zero Z 2.0. And after that we will do tool homing. So here we have completed our first drilling. That means that for M drilling. Now we will do second drilling and diameter of that drill H6MM. So for this I will write n6. Then we will call our 6MM drill. After that we rapidly call it up to Z 2.0. And we have to do this 6MM drilling up to -20 as our this distance is 20 m m. So we have to do this drilling. Similarly, as we have done for for m m. So firstly take occurred in Z up to -5.0. Then take that tool back here I have taken large cuts. But you have to take small cuts only then Z01Z -10.0. Then again back. Then again take cut. Then again back to backward. Process is with G00 code and cutting process is with G01 code. So this you already know. So for last cut we will go Z01Z -20.0. And then we will take that tool out of the job up to Z 2.0. And then we will do tool homing. And now we will write M05M09 and M30. End of program. So here our total job program is completed. So that's how you have to write program to make this type of job from raw material. So here we have covered all the operations. And also we made program for our job from raw material. So now we will do one thing. We will run this last example in our Fanuc as well as NCM as control. So that you will understand this much better. So firstly we will run this program in FANUC control. So as you can see here, I have written already program here in FANUC control. Exactly same program I have written here. Now I am running this program in single block. So see you only have to compare your written program with this actual tool movement. So it is working very accurately. So see this extra material. Remove exactly as per our program. In both the places. So this was roughing process. Now see, it is doing finishing process. So see finishing process completed. After that we will do threading. So this tool will do threading. Here. Actually in this simulator I was not able to find this 1.5 page insert. And that is why this inside is colliding with job. But if you use proper size insert then you will not get this problem as this is G 92 cycle. So tool will lift up in this direction. And that is why it is touching our job. You only have to see whether this G 92 code is making threads or not. And as you can see here, this code is making threads in both the places after that groove. You only have to see whether that groove is actually making or not. By our program I also see its cutting direction and whether that cutting direction is exactly same as per our program or not. So it is exactly working as per our program. Now you can see drill compare your written program with actual tool movement, whether that cutting is exactly as per our program or not. Here in this simulator, we can only use one drill at a time. So that is why I have shown you this drilling by using only one drill. Now we will check its dimensions as well so that we can confirm that our job is dimensionally correct. So see our miner diameter of this thread. This was 8.7 ten.

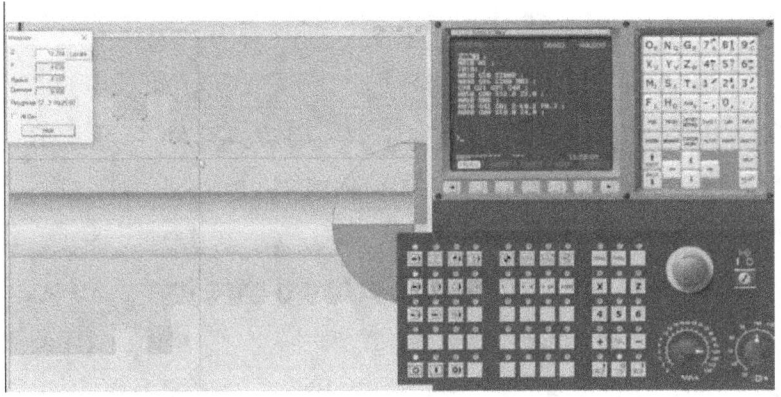

So see whether this miner diameter is 8.17 or 0. So we are getting nearly same dimension. Now we will see this diameter. So this diameter was 16. So see we are getting nearly 16 diameter. And also check its length. Here in Z it was 18.5. So we are getting nearly 18.5. After that we will check this length. It was -30. So see we are getting nearly 30 distance. So now we will check miner diameter of this thread. So miner diameter of this thread was 17.865. So see here. So we are getting nearly 17.865 diameter. Then check this length as well. It was -50. So we are getting nearly same length. This diameter was 30. So see now we will do drill measurement. So we have done drilling up to Z -60. So see we are getting nearly same dimension. So we get accurate dimensions for this job. That means our program is correct. So this is all about FANUC control. Now we will see this program in Siemens control. Whether it is actual working in Siemens control or not. So here also I have written all the program. You can see this program. I have written everything exactly as per our program. Firstly, roughing. And then finishing everything I have written here. In Siemens training cycle, it's somewhat different. We can't use G 92 cycle here. Here we get form for cycle and we have to fill that form. So we will see this in our cycle topic. So don't worry about this cycle. And remaining all program is same. I have written here drilling and all everything I have written here. So let's simulate it. So firstly we will have roughing then finishing. So all this process will take place one by one. So see firstly it is doing roughing. After that finishing. Then threading. So see here we do not get any alarm as we have used proper thread insert on insert for each value is written. So always use insert equal to thread page after that grooving. And after that drilling. So our all operations are covered here. And we get job as per our requirement. That means our program is accurate here. Everything is covered here. Drilling grooving. Everything. So here we have seen one example with all the operations. And we run that program in Fanuc as well as in Siemens. So here we have completed our example topic.

CYCLES (PART 1)

So in last topic, we have seen many examples and we write program for that by using g zero, zero and G01 code. And there we have taken large curves. So if you write that program with small curves then you will require nearly 4 to 5 pages to write that program. And instead of that G00 and G01 code, if you make that same program with cycles, then you will require only one page to write that whole program. That means we can save over time as well as efforts with these cycles. So in cycle you only have to write two three lines of that cycle and then write your finishing program. That's it. How simple it is. So let us see these cycles one by one. So our first cycle is stock removal. Cycle means to remove extra material from raw workpiece. We use this cycle. So in this G70 and G70 one needs two codes. So G70 one code is for a roughing cycle means the roughing cycle of stock removal cycle. So G70 one will do roughing and G70 will do finishing. So let's see what's in our roughing cycle means Z71 cycle. So G70 one roughing cycle then you. So this you each depth of cut in x means how much depth of cut we are going to take in x direction. That value we have to give here. R means lift after cut means how much distance that tool will lift after taking cut that value. We have to give here. So this is our first line of our cycle. And our second line is G70 1PB. Each start block number means after these two lines we are going to write finishing cut means finishing program. So this G70 one cycle rate that finishing good. And according to that finishing curves it makes its roughing cuts itself. So you only have to give starting block of our finishing cut. So we write that starting block in P in queue we write end block of our finishing cut. In you we write finishing allowance in X. That means how much material we are going to keep in X for finishing that we have to write here W is finishing allowance in Z. That means how much material we are going to keep in Z that we have to write here. And this feed that you already know. So here I have written g 71 U1 means depth of cut is 1MR means left. So after cut I'm going to leave that tool by 1MM starting block of my finishing cut is 40 and end block of my finishing cut is 110. I have given finishing allowance in X as 0.5 M in Z. I have given 0.3 and feed 0.3. So this is our roughing cycle here P 40 and Q 110. This p and q values we write in last means. Firstly we write finishing program and then only we can write this p and q p means starting block number of our finishing program. And q is end block of our finishing program. We can't write these numbers before writing our finishing program. So this is our roughing cycle. Now we will see G70 cycle means finishing cycle of stock removal cycle. So in this we have only two parameters. First is b and second is q. P is starting block number of our finishing program and q means and block number. Now let's see one example so that you can understand this cycle very well. So we will write some lines of our standard format file name tool homing then tool call. Then we will give some instructions to our spindle after that rule line. And then we will rapidly call our tool. So here diameter of our drawing is 140 and support our raw material. Diameter is 145. So firstly I will take a small cut. So I will call my tool at x 140 2.0 and z 2.0. And then I will take occurred here in Z up to -170 as my total job length is 170. So I will write z minus one 70.0. And then I will lift that tool up to 200 and z 2.0. And after this we will write our cycle. So as you can see here these N10 and 20 and 30 we do not have to write these codes. Whenever you give end of block and then press insert. This next block number will automatically come means this N10. Then machine will write and 20. This process is automatic. We don't have to write these n numbers. So we were in G 71 cycle. So first line of our z71 cycle eight G 71. You and R. So you need depth of cut. So here we have taken depth of good adds 1MM and are its lift after cut means after cutting how much distance we have to lift our tool. So here we have written one m means tool will lift by 1MM. Our next line was G £0.71 means starting block. So here I have written finishing program of our job. So starting block of our finishing program 890 and end block is 170. So you have to write these numbers in P and q. After that you are finishing allowance in x.

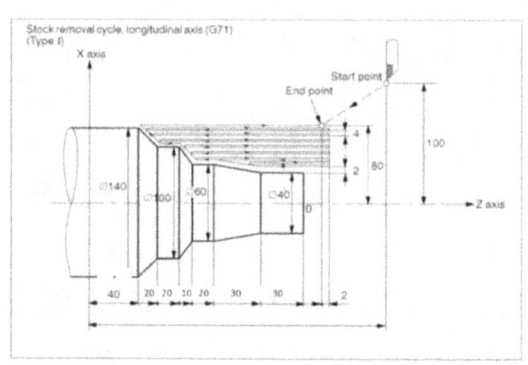

```
O1234
N10 G28 U0.0 W0.0;
N20 T0101;
N30 G50 S2000;
N40 G96 S160 M03;
N50 G90 G21 G95 G40;
G00 X142.0 Z2.0;
G01 Z-170.0 F0.3;
N60 G00 X200.0 Z2.0 M08;
N70 G71 U1 R1;
N80 G71 P90 Q170 U0.5 W0.3 F0.3;
N90 G01 G42 X0.0 Z0.0 F0.3;
N100 G01 X40.0 ;
N110 G01 Z-30.0;
N120 G01 X60.0 Z-60.0;
N130 G01 Z-80.0;
N140 G01 X100.0 Z-90.0;
N150 G01 Z-110.0;
N160 G01 X140.0 Z-130.0;
N170 G01 Z-170.0;
```

So I have given 0.5 finishing allowance in Z 0.3 and feed 0.3. Now we will see this finishing program. So firstly we will touch the tool lightly to the face of the job. So we will write G01G 42 x 0.0 z 0.0. And feed 0.3 means we will lightly touch our tool to the jaw, face. After that we will write G01X 40.0 means tool will take it from this two. This. After that G01Z -30.0 as this distance is 30. So we have written z -30.0. After that this step. And for that we will write G01X 60.0 z -60.0. And after this we have to take a straight cut. So we will write G01Z -80.0. This distance is 80. After that once again we have this taper. So we will write G01X 100.0 z -90.0. After that one second we have a straight line. So we will write G01Z -110 0.0. After that once again we have this taper. So we will write G01X1 40.0 z minus one 30.0. And then in last we have a straight cut. So we will write G01Z minus one 70.0. So this is our finishing program. So CNC machine will fit this program in its mind. And according to that it will make roughing cuts by itself in G 71 cycle. And according to that tool will cut the material means by considering this finishing cycle CNC will make a roughing cuts. So here we have completed our roughing. Now we have 0.5 and 0.3 allowance in our G 71 cycle means we have to cut this allowance material as well. And for that we will write finishing cycle as well. So we will write G70 Finishing cycle. So G70 be our starting block. It is 70 and Q 170 and our block. And then we will write T0404 at zero four is our finishing tool means we will call our finishing tool. And then we will do this finishing cut. So you only have to give this information machine will do rest of the work by itself. After that lift that tool. So we will write G00X 200 and z 2.0. And after that M05, M09 and m 30. So this is how you have to write this torque removal cycle. So see how simple it is. We have converted our 4 to 5 page program into this simple program by using cycles. That means this cycle saves our lots of time as well as efforts. So you only have to write these two lines of roughing cycle and this one line of finishing cycle. And between these cycles we have to write our finishing program. So since the machine will make its roughing cuts according to this finishing program and do roughing as well as finishing. So now we will write this program in final control and we will run it. So here I have already written all the program so that we can see over time. So I have written everything here. Two lines of our Z71 cycle. And I have also written G70 cycle after finishing program. So program that we have recently seen that same program I have written here. So now let's run this program in single block so that you can understand much better that how our cycle works. So just see this. So firstly we will take a cut in Z. And after that we will start our stock removal cycle. So see our tool is cutting material. According to G70 one cycle means machine have read already finishing program. And with that program in its mind. And according to that machine have made its roughing cuts. And with that roughing cuts, our two leads working now.

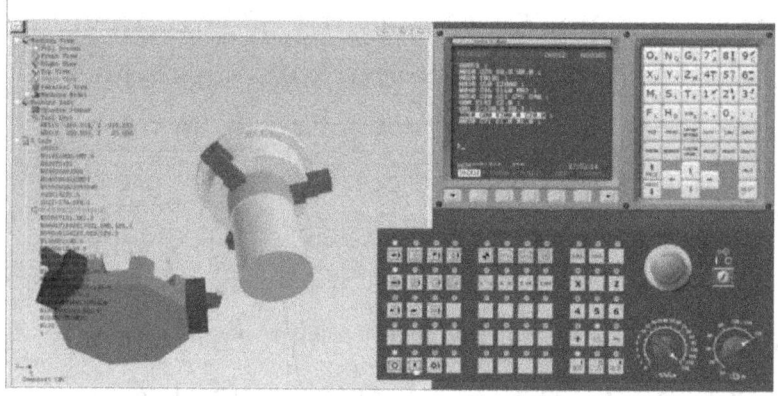

So see how simple it is? How simple. They made our program. Now it is taking one light cut. This is not finishing. This is roughing. Only tool is taking light cut in roughing. So that there will be equal material everywhere for finishing. So, see, this cycle made our program very easy. We save our lots of time and efforts here. You only have to write, cycle and then write. Finishing program. That's it. CNC machine will do. Remaining all work by itself. So this roughing. Now we will see. Finishing. See tool change. And now that tool will do finishing. It will read one line, run it and then stop. As we are running this in single block. So see, our job is ready. How simple it is. So we have seen this torque removal cycle in Fanuc control. Now we will see this cycle in Siemens control. So now we will see this simple example by stock removal cycle. And after this we will see counter stock removal cycle means in that we will have taper and radius. So firstly we will see this simple example in Siemens control. So until now I have not told you how to make new program in Siemens control. So in this project I'm going to tell you that how to make a new program in Siemens control. And we will see these simple example with stock removal cycle. So we are going to see all this in Siemens control. So firstly we have to press Program manager button to make new program press program manager button. So in that you will find part program F sub programs and workpieces. So what is this part program means if we want to make new program then you have to select part program. Sub programs are selected when any part of our program is repeating again. And again that repeating part. We write in sub program. And whenever we want that sub program in our main program then we just call that sub program. So this is mostly use in VMC machine. We rarely use this in class in machine as we have cycles in CNC. So we don't have that much use of these sub programs. But in VMC some small process keeps repeating again and again. So we write that small program in sub program and then we just call that sub program in our main program. So all these sub programs are written here.

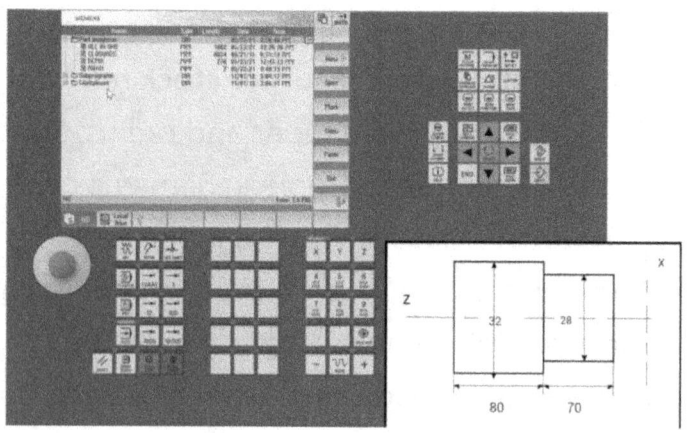

And here we select workpieces means in some three months we have to give workpiece information to machine. So that workpiece information we write here. So now we have to make a new program. So we will go in part program. And here we will select new. And as I have already told you that in Siemens you can give any name to your program. We don't have any rule like final control.

So we will write program name adds simple stop cycle. And after writing this name you have to press okay. So our new program is open here. And now we have to write first some lines of our standard format. So we will write first some lines of our standard format as they are common in all programs. So one then we will give some spindle instructions here. I will give all instructions in one line only. After that rule line. So we can interchange first four five lines of our standard format, as that will not affect that much on our program. So in some Siemens we have to give workpiece information. And in some Siemens we do not need this. So I will tell you how to give this work this information. So fourth is firstly select various option. And in various option you have to select this blank option. So this window will open. And here we have to give our workpiece information to Siemens control. So firstly we have blank means which type of raw material we are going to select here. So we are going to use cylindrical material. So we will give this blank add cylinder. After that x a means diameter of our raw material. So we have to write this diameter here. So now our example have 32 diameter. So we will take raw material as 40 diameter. After that z a means at this point we create zero zero. So this is asking whether you have to keep this point at zero zero or you want to give some other dimension here. So according to that machine we will do remaining calculations. But we create this point at zero zero means in setting we create this point at job zero point, this center point. So we will keep this z a at zero zero. After that we have z one means we have to give here length of our workpiece. So we will take 200 length as our example length h 150. So we will take 200 length. And z b means chunk distance from z. It means this chunk distance from job zero point. Here I don't want to show this jerk. So I will write here -200. But you have to measure this juggling from job zero. And then write that value here. And after writing all these values, simply press accept key. So that workpiece will be added in our main program. And after writing all these we have to do tool homing. So we will do tool homing. And after that we have to call our tool. So in some Siemens control you have to tell Siemens each cell that which tool we are going to use. And in some Siemens you can directly call that tool just by writing T01D1. So we will see both methods. So firstly we will tell Siemens itself about tool that we are going to use. So for that press edit after that select tool. So we are going to select our tool from here we have to select our tool from this chart. So firstly we will choose Roughing Tool and then press okay. So that tool will add here. And after that we will write d1. This D1 is to cancel all the preceding effects present in that tool. And our second method is we can directly call our tool. So we will write D01D1. So you can use any one from these two methods according to your machine. So after writing all these we have to call our tool rapidly near our job press. So we will call at x 42 and z 2.0. So we have rapidly call our tool. And after that we are going to see this simple stock removal cycle. So for that we will go in turning. So this is turning option. You have to select this option for our simple stock removal cycle. And in turning you have to select stock removal option. So this window will open. So in this window we have to fill all the information that they have asked. So first it's C means safety distance or safety clearance means we have to give some safety distance in x and z. So we will write here 1MM. So two tool will stand one m outside that required position.

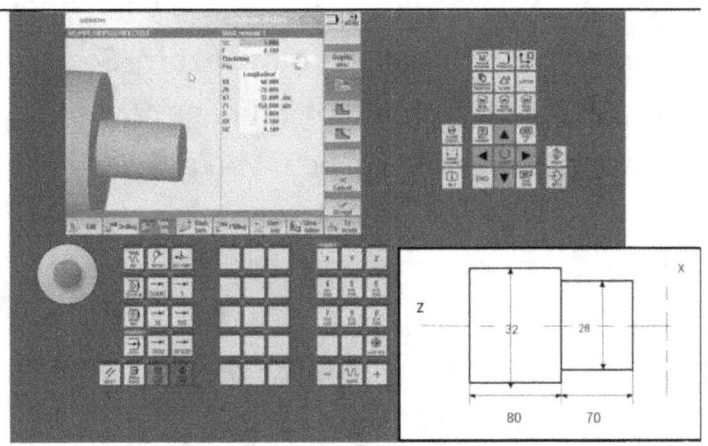

When we rapidly call it. After that we have freed. So now you know this field. After that we have machining. So what are you going to perform? Either roughing means if there is only one triangle then it is for roughing. And three triangles is for finishing. So here firstly we are going to do roughing. So we will select one triangle. After that we have machining. Direction means in which direction you are going to cut this material. So here we have two options for that. First is longitudinal means. It will take cut like this and then lift it. As you can see in animation. So this is how it is going to cut material and how tool will cut material in phase. So see this with the help of animation as well. So this is how it is going to cut material. So we will choose longitudinal here. After that we have to give x zero and z zero. Value means this point that they have shown us that points dimensions. We have to write here in x zero and z zero. So our raw material is of 40 diameter. And that is why our x zero will come 40. And what about z zero. So we have created starting point at zero. So we will write here zero zero. After that x one means how much diameter we want to make here. So if you can see in our drawing.

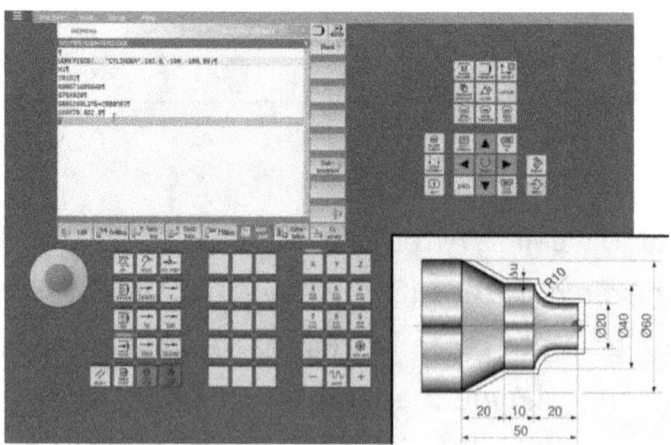

So we have to make 28 diameter here. And that is why we will write 28 in x one. And these values are in absolute. We have already learned this absolute mode and increment mode. So we have to select absolute mode here. And how much this z one will come. So if you see length in our drawing so that length is 70. So that's why we will write here -70. After that d means depth of cut. So we will take one more depth of cut. So we will write one in D after that. UX and use z means finishing allowance in X and finishing allowance in Z. So we will write this allowance at 0.1 in x and 0.21 in Z. And after writing all these values, we have to press accept key. So here we tell in machine that how we are going to cut this material. Now we have to make stock removal roughing cycle for 32 diameter. So let's make that cycle as well. So for that select turning option. And in that turning option select stock removal option. After that you have to fill this form one second. So SC means safety clearance. That will be same feed also same machining also same. And we will keep our machining direction at longitudinal. Now x zero means our this point will shift here. So it will shift here. So how much will be diameter at this point. So we will have 40 diameter here as well as our raw material is our 40 diameter. And what about z distance here. So z distance will come -70. So we will write -70 here. Now x one means how much diameter we want here. So here we are going to make 32 diameter. So we will write here 32. After that Z one as we are taking this length in absolute.

So we have to take here total length. So as you can see here our total length is 150. So we will write here -150. After that depth of cut one m and UX and you z. So we will take this 0.1 allowance in x and z. And after that press accept key. So we have created roughing cycle for 32 as well. So after that we will do tool homing. So up to here we have completed our roughing cycle. Now we will do finishing. So for that we will write and tool. And then we will call our finishing tool. So I will directly ride here T03D1 as my offset is in one number offset. As in this simulator they have given only one option for giving offset. And that's why we are writing here T03D1. But you have to write T03D3 as you will write three number tools offset in three number or wherever you are going to write that offset. So that number you have to write after D. After that we will go in turning and in turning option. Now we will do finishing as we have completed our roughing. So S, C and FID will be same. And as we are going to do finishing here. So we will select three triangles in machining option after that longitudinal after that zero. So we have to write dimensions of this point in zero and z zero. So write that dimensions for 28 as. Firstly we are going to do finishing word 28. After that x one means how much diameter we want here. So we want 28 diameter. And z one will be -70 as this length is 70. And after that press exit key. So this finishing cycle will add here as well. So we have made finishing cycle for 28 diameter. Now we will make finishing cycle for 32 diameter as well. So again go in turning then in stock removal. And then fill this form. So here x zero will be same 40. This point will shift here. So z zero will be -70. Then x one. So how much value will come here. So x one will be 32 and z one will be -150. So these values we are taking in absolute mode means all these values will be measured from job zero point. And after that press accept key means. We have completed our finishing cycle as well. So that is how we have to write roughing as well as finishing cycle. For this type of simple job. And after that right. And 32 and that program. Now we will run this program and we will see whether it actually works or not. So let us simulated. So firstly it will do roughing process. Now it is doing roughing for 28 diameter. After that it will do roughing for 32. So see it is doing roughing for 32 as well. Now it will start finishing. Okay. Means our cycles are working properly. So that is how we run this stock removal cycle in value control as well as in Siemens control. But in Siemens control we have taken very simple job. So now we will see counter stock removal cycle means how to make this radius and chamfer in this cycle. So we will see that as well. We have little difference in it. That is why I'm covering this as well. So firstly we have to go in part program. And there we have to create a new program. So here we will write name of our program add counter stock removal after that okay. So our new program is created. So firstly we have to add workpiece here. Now you know how to add this workpiece. Firstly press various key then blank then fill this data. So here we are taking diameter at 65. And z will be zero and then z one. So we will take length at -100. And we will write z be at -100 as well. Now you know all these parameters. So our workpiece will add here. After that we will write first for fine lines of our standard format. So here I have called roughing tool as well. So we will write these four file lines. Firstly. After that we will rapidly call our tool. So our raw material is of diameter 65. So we will call our tool at X 70.0. And we have to make cycle means we have to make our counter stock removal cycle. So for that we have to select counter turning option. And in that option we have to select counter call option. Firstly we will call our counter. We will make this counter later. But firstly we will call it. So in this counter name we have to write name of our counter that we are going to make later. So you have to remember this name as we are going to use this name later. So here we will write name as counter. So remember this name and then accept. So here we called our counter program that we are going to make we make this program later. But firstly we call it. After that we have to give some parameters to our cutting tool. Means in which direction our tool have to work, how much depth of cut it have to take. So we will give all these parameters to our tool and to give that parameters we have to select counter turning option. And in counted turning option we have to select stock removal. So you will get this form. So you have to fill all the information here. Firstly we have input complete. So we will keep this complete as we are going to make complete program after that program name. So you can give any name to your program here. So I will give name as ABCd and E. You can give any name after that. See safety clearance. Now you know this safety clearance then you also know feed. After that machining one triangle plus three triangle means it will do roughing as well as finishing. So here we are going to perform both the operations. And that is why we select this option. And after this we will write finishing feed as 0.1. After that longitudinal. And after that outside and outside means body turning. Inside means returning. So here we are going to take cut in order. So we will select outside option here. After that depth of cut. So we will keep this depth of cut as one m after that UX and use z allowances. So we will give that allowances at 0.1. After that d I means if you want to take cut in this way means if we give any value here, let's suppose we give here value as two. So tool will take cut of two m then move back that it will add two more in that. And again take cut. And again it will move back. Again it will add two more. Again it will take cut and again it will move back. So if you want to take this type of cut then give value here. But we don't want this type of cut. We will take this type of straight cut. And that is why we will write here zero zero after that blank. So blank. It's cylindrical. So we will keep this cylinder as it is. After that x d and z d means how much allowance in x and how much allowance in z you have to give. You have to remember that give these allowances in increment only means how much allowance you want above our diameter. So we want one more allowance from our diameter. That is why we will write what m in and c in both z and z. After that, relief cuts means after taking cut. Whether you have to lift that tool or not. So select yes and then write value here. Moving on our next parameter h limit. So if you want to add any limit then you can add here. This is rarely used means this limit is given in special condition. So select no here. And after that press accept. So all the tool instructions will add here in cycle format. After that we will lift our tool up to x 100. And then we will do two roaming. And then M 30 here we have not completed our program. We have called our counter program. That counter program we have to write and we will write that program after M 30. And that is why we will write our counter name program after M 30. So let us start to write that program. So first select counter turning. And in that option select counter option. And in that option select new Counter option. So here it will us to enter name. So here we have to give name which we have called in our program. So we have called counter. So here we have to write counter whichever program we are calling that must be present here. If you call program by another name and you write program with another name. So that is not going to work. In that case program will call. But as there is no program present of that name so nothing will run. So always give name of your program that you have called. So here we will write counter and then accept. So this window will open. Now we have to make profile here means according to our job profile we make profile here. So our first point is x 20 and z 0.0. So we will write here that value x 20 and z 0.0. And after that accept. So we will get this point. After that we have stretched horizontal line. And what is its length. So if you can see in diagram. So this straight line is up to z minus ten. So we will draw a straight line up to minus ten means up to z minus ten. So we will write minus ten in z. After this straight line we have a curve here. So we will make that go as well. So select this option and in which direction we have that curve. So we have curve in clockwise direction. So we will select this clockwise curve. After that we have to give radius for it. So we have radius of ten x will be 40. As you can see in diagram and z distance is up to -20. So machine will do all the calculations and it will show two radius curve. So you have to select curve that you want from these two curves. So here we don't want that selected curve. We want another curve. And for that we will press dialog select so that our required curve will get selected. And after that press dialog accept key. So machine will automatically do its calculation for that curve. And then press accept button. So our required curve will appear here. After that we have straight line. So we will draw a straight line here. So we will select straight line. And then we will write z -30. You can see these values in our diagram okay. And then accept. After that we have table. And to draw this table we will select this option. And then we will write x and z values. So our x is 60 and z is -50. So write these values in x and z. So machine will do all the calculations by itself and then press accept button. So this stepper also gets appear here. After that we have a straight line and length of that straight line from job zero is -70. So we will write -70 and then accept. So profile of our job is ready after that. Once again press accept. So all the course of our profile will come here. So let's see how this program is going to work. So firstly we have added our workpiece. After that we have written 4 or 5 lines of our program as per our standard format. And these lines are almost same for all the programs. After that we called Counter program. So machine will read this profile, fit that profile in its mind and then it will look in our second cycle that how much could it have to take its cutting direction and all the other parameters. And according to that machine will run that cycle. And in last it will lift the tool. And then the tool homing. So in this way this is going to work. So let's simulate it and we will see whether it runs or not. So see he had same tool will do roughing as well as finishing process. As we have used only one tool. So that single tool will do both the operations. So machine have made same profile or job as we want. Means our cycle is correct. So in this way you have to make counter stock removal cycle. So here we have seen stock removal cycle in Fanuc control as well as in Siemens control. So now we will see our next cycle. So our next cycle is pattern repeating cycle. And for this cycle we use 73 code. So many times what happens you will get a forging job or casting job. So if you do forging or casting in that case there will be only few material present means there will be only few material present above our required job. So in that case, if we use G 71 cycle, then there will be wastage of our time. As in many passes, there will not be any cutting of material as we have only 4 to 5 M material present. There means you only have to remove these 4 to 5 M material remaining. All material is already removed in forging or casting. So you cannot use G 71 cycle. So in this condition we have G 73 cycle means whenever you are job is having only 4 to 5 M extra material available everywhere. So in that case youth G 73 cycle means pattern repeating cycle. So in this pattern repeating cycle G 73 is roughing cycle and G 78 finishing cycle. As you already know this. So let us see G 73 cycle. So in this cycle we have. You mean stock in X means how much stock is present in x direction that you have to major. And right here w each stock in Z means how much extra material is there in Z that you have to write here. And r is number of passes. So let's suppose if you have five m extra material in x and z direction and you are going to take one, then depth of cut means in every pass you are going to cut one m material. So to remove 5MM material you will require five passes. So that value or number you have to write here.

Pattern Repeating Cycle (G73)
(Mainly used for forging or casting job)

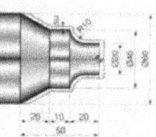

(G73----Roughing Cycle)

G73 U_ W_ R_

G73 P_ Q_ U_W_F_

(First Block)
U---- Stock in X
W----Stock in Z
R----Number of passes

(Second Block)
P----Cycle start block
Q----Cycle end block
U----Finishing allowance for X
W----Finishing allowance for Z
F----- Feed

After that we have p means cycle start block. So this we have already seen in one and G70 cycle. There we had written cycle. And after that finishing program. So starting block of our finishing program is in P. And last block of that finishing program is in Q. What will be in you? Finishing allowance in X means how much material we have to keep for finishing in X that we have to write here. W means finishing allowance in Z means how much material you have to keep in Z that you have to write here. And this feed that you already know. Sometimes there may be condition that you have to make this type of job from raw material with the help of G 73 cycle. Normally for such types of jobs we use G 71 and G 70 cycle. But if in special case there is need to use G 73 cycle for that raw material, then let us see how to do this. So in this cycle, as our first line of G 73 cycle, each stock in X stock in Z and number of passes. So we will learn here how to calculate these three parameters. If you have forging job then you can get these values directly by measurement. But here as we have raw material and from that raw material we have to make this diagram. So we have to calculate this. You and R. So to find stock in x we have one formula. So you stock in X is equal to x is material diameter divided by two. So what is this axis. Material diameter. So it is diameter of raw material minus smallest diameter of job. So what is diameter of our raw material. So let's suppose we have taken raw material diameter as 145 m. And in that we will take one cut up to x 140 2.0. So our raw material will become 142 diameter. After that what is the smallest diameter 40. You can see here smallest diameter is 40. So if we do this calculation then we will get excess material diameter at 102. So we will put this value in our you formula means in stock in egg formula. So we will get value of u at 51 m means 51 M is our stock in x, W means stock in Z. So you can take any stock in Z or just major, this stock in Z in your job and number of passes. So let's suppose I want to take one m depth of cut in every path. So to cut 51 m material we have to take 51 passes. So in this way you have to find three parameters. And this we will find when we have a raw material. And from that raw material, if we want to make this type of job using G 73 cycle, then only we need this formula to find out these three parameters. If you have forging or casting job, in that case, you will get values for these parameters by measurement. But if you have raw material and from that raw material, you want to make this type of job with G 73 cycle, then only you have to use this formula. Normally we do this type of jobs with G 71 and G 70 cycle. But in special condition. If there is need to use G 73 cycle, then you have to use these three formulas. So this is G7 D three roughing cycle of pattern repeating cycle. And to remove allowances given in roughing cycle we use finishing cycle means G7 cycle. So this is finishing cycle. So in G 70 we have p and q starting block number and end block number. So let's see how to make program for it. So here we will make this job from raw material with G 73 cycle. So as per our standard format we will write our first five six lines which are common in all the programs. Then we will rapidly call our tool up to X 140 2.0. Then we will take a cut in Z up to - 170 as our total length is 170. So we will write z -170. After that we will lift our tool x 200 and z 2.0. And now we will write our g 73. Cycle means pattern repeating cycle. So first line of our G 73 cycle is used w and r. So we have already calculated these values u is 51 means stock in x which is 51. We have taken stock in Z at 0.2 m and number of passes is equal to 51 means we are going to take depth of cut as one m and in second line we have p means starting block of our finishing cycle, which is 90, and block is 170. After that you finish allowance in X which is 0.5, finishing allowance in Z within 0.3 and fade. So we have taken feed at 0.3. And now we will write. This finishing cycle means this finishing program. So firstly does the tool do the job phase G01G 42 x 0.0, Z 0.0 and feed 0.3. Then we will take it in x up to x 40, then Z01, z -30.0. Then we will take cut up to x 60.0, z -60.0. After that Z01Z -80.0. After that Z01X 100.0, z -90.0. Then Z01Z -110.0. After that Z01X1 40.00, minus one 30.0 and in last Z01Z minus one 70.0. So we write this finishing program here. And now we will write our finishing cycle G £0.70 and Q and call our finishing tool as well. After that we will lift our tool and then M05 and zero nine and M30. So in this way we write this G 73 cycle means pattern repeating cycle. Now let us see this with the help of project and we will see whether we get our required job from raw material or not. So here I have written all the codes I have written G 73 cycle. Here I have reduced number of passes and as 51 passes will require more time. And I have also written G70 cycle. So in this project you have to see how this G 73 cycle works means how our tool is cutting the material. And you have to compare this with our G 71 cycle. I'm telling once again, this G 73 cycle is used for forging or casting jobs. If you have to make this job from raw material, then use G 71 cycle. As in this simulator we do not have forging job. So we are using G 73 cycle on raw material. So when you use G 73 cycle on forging job, then you will get u w and other values means stock in x, stock in z and number of passes value directly by measurement. There is no need to do any calculation for these values. So in this project you only have to see the cutting direction of g 73 cycle. So by looking at the cutting direction of G 73 cycle, you will know why we use G 73 cycle for forging or casting job. So see why we call g 73 cycle as pattern repeating cycle. Look at this cutting direction. It is cutting material in pattern. So pattern repeating cycle. So as you can see here when we use G 73 cycle on raw material then there is wastage of time. As in many curves tool is not removing any material. Tool is taking only small portion of cut and remaining. All tool movement is wasted. And that is why this G 73 cycle is used for forging or casting job. You have to use G 71 cycle in such conditions. Means when you have a raw material then you have to use G 71 cycle instead of G 73 cycle. So, as you can see, it is running one line only, as we have done on this single block. So our job is ready. So we have made this job by G 73 cycle as well as G 71 cycle. I hope you understand the difference between these two cycles. G 73 cycle is used for forging or casting jobs, and G 71 cycle is used for making job from raw material. So this is a pattern repeating cycle in fan control. Now we will see this cycle in Siemens control. So we will make this type of job in Siemens control with pattern repeating cycle. So see firstly we will make a new program here in part program. So this I have already told you. So here we will write name as pattern. Firstly we will add our workpiece here. So we will take 65 diameter workpiece. And we will keep this length 100. As it is. After that we will write four fine lines according to our standard format. After that, we will rapidly call our tool. And after that, we are going to make pattern repeating cycle. And we will make this cycle by doing some modification in counter stock removal cycle means. Firstly, we will make finishing program of counter stock removal cycle. And after that we will do some modification in it to get our pattern repeating cycle. So let's see. So firstly select counter turning option. After that select counter and then counter call. So you have to remember name that you give here. After that press accept. So we called our program. After that again select Counter Turning. And in that counter turning now select stock Removal option. So we will get this form. So you have to fill this form. So we will keep this input as complete as you know safety clearance. Then feed. And in machining you have to select finishing. Remember it when you are going to make pattern repeating cycle at that time you have to select finishing. Here. After that everything will be same longitudinal. Outside you have to select allowance as no. We will give that allowance later. So keep it. No remaining. All is same and give name to this. You can use any name here. So give any name here. And then press accept. So all our cutting parameters will appear here in cycle format. After that leave that tool. Then do tool homing. And then write m30. And after that we will write program that we have called means this pattern r e. We will make that program here after M30. So for that again press counter turning option in that select counter and then select New Counter option. And here we have to write name of program that we have called. So we called pattern R. So we write name here as pattern RSP and then press accept key. So after that we have to make here profile of our job. So let's make this profile. Our first point is x 20 and z 0.0. After that we have a straight line here. So we will write z -10.0. After that we have come. So we will make this curve. Here we have clockwise curve with radius ten. And x is 40 and z is -20. As you can see in diagram. So select curve that your profile have from these two curves. And then press accept. Now we have straight line. So we will write z -30 as our straight line is up to -30. After that we have tapered line. So our x is 60 and z is -50. So our taper line will also appear here. After that we have straight line. So we will write z -70.0. And then we will press accept. So our profile is ready here. So let's understand the flow of our program. So firstly it will call program of name pattern r e p. So it will find that program. And then CNC machine will read that pattern and read it in its mind. Then machine will see tool cutting parameters means depth of cut allowance. Everything machine will see here. And according to that tool will take cut. But here we have only created a finishing cut of counter stock removal cycle. But we want pattern repeating cycle. So let's see how to do this. So let's suppose we have three more stock available on our job. So we have to remove these three material by taking one more depth of cut means tool will take three cuts each with one more depth of cut. So we have to make three passes here. So for that the cursor on our cycle. And then press this key so that our cycle will open here that we have already created. So in that what do you have to do. You have to change allowance from no to yes. And in Q1 we will write 2.0 M allowance means as we know we have three m material available. So when we take first cut of one m then how much material will remain three minus one which is equal to two means two m material will remain there. And that's why we have written 2.0 here. And then press accept. Now we have to copy this cycle. So for that we will select edit and then copy. And we will paste this cycle here only as we are going to take three cuts. So we will paste it for three times. So we have created three cycles here. So in our first cycle we have given two m allowance. So now we have to select second cycle. Then we will open it. And in that we have to give allowance as 1MM means when tool will take second cut. Then how much material will remain there. Means firstly we have three m material, then we take first cut off 1MM. So two m material remainder that we have written in our first cycle. And now we are taking second cut. So one material will remain there. So that's why we will write here 1.0 and then press accept. And in third cycle how much material will remain there after cut. So there will not be any material remain. And that is why we will select allowance as no means. This is our finishing cut. So that is how we have to make this pattern repeating cycle. You have to copy paste these cycles according to material available on your job, and whatever stock will remain that you have to write in allowance. So here we have completed our pattern repeating cycle program. So firstly it will call pattern RSP program. Then it will run that profile and fit that profile in its mind. And according to that it will take first cut means it will keep two M allowance. After that it will keep one m allowance and take cut. And after that it will take final cut means it is taking cut in pattern. So now let us see how this program actually works. So let us simulated. Here tool have taken cut in this way as we have only three material available. And that is why tool have taken one more cut in this way. And now two will have to take two more cuts. So two will take two more cuts. So here you will find three cuts means each time two will have taken, 1 a.m. cut. So in this way we have to make this pattern repeating cycle in Siemens control means firstly make finishing cut of counter stock removal cycle and then reduce allowance by one. Every time means copy that cycle, paste it and reduce allowance by one m. So if you have y m stop available on your job, then you have to copy that cycle five times and reduce that allowance by one m every time. And in last we have to select allowance as no means that cut will be finishing cut. So in this way we have to make pattern repeating cycle in Siemens control. So here we have seen this pattern repeating cycle in Fanuc control as well as Siemens control. So until now we have seen two cycles. First what stock removal cycle G 71 and

G 70. And second if pattern repeating cycle G 73 and G 70. So we have seen this in Siemens as well as in. So we will see our remaining cycles in our next part.

CYCLES (PART 2)

So in last topic we have seen two cycles. First its stock removal cycle and second each pattern repeating cycle. So let's see our next cycles. So our third cycle is drill cycle. And for this cycle we use G7 for code. So in this cycle also you will get two blocks. So in our first block we have G 74 and R. So these are each retraction amount. So if you know how this drilling works it do not take whole cut in one time. First two will take light cut then it move back. Then again it take good and again move back. So in this way tool makes that drill. So that backward distance is known as retraction amount. That means tool will take good and then move back. So that backward distance we write here. And in second block we write g 74 x. So this x is diameter up to which drilling has to be done means up to which diameter we are going to take. Good. But we don't do. Drilling in diameter means we don't do drilling in X direction. We do drilling in Z direction. So that is why we don't write this x in drill cycle. After that we have z means distance up to which drilling has to be done, means up to how much distance we are going to make that drill. So that z distance we have to write here, be it depth of cut in x axis. But as we are not going to take good in x axis, that's why we do not write this p in drilling cycle as we do grooving in x axis and drilling in z axis. That is why we don't write this p here. After that q depth of cut in z axis. So as we do drilling in z axis. So here we have to write depth of cut and that depth of cut. We have to write in microns. So I have already told you this m m to micron convergence. So here you have to convert depth of good in microns. And that value you have to write here.

Drill Cycle(G74)

G74 R_

G74 X_ Z_ P_ Q_ F_

(0.1*1000 = 100)

(We have used 40mm drill)

(First Block)
R----Retraction amount

(Second Block)
X----Diameter upto which drilling has to be done
Z----Distance upto which drilling has to be done
P----Depth of cut in X axis
Q----Depth of cut in Z-axis (in microns)
F----Feed

So suppose we are going to take 0.1 m depth of cut. And to convert M into microns we have to multiply it by 1000. So if we multiply 0.1 m by 1000 then we get 100 microns. So you have to write 100 here. And after that we have feed. So this feed you already know. So here we have used for t m drill. In drill offset we have a little change. So let's see that change as well. So firstly we have to touch D drill to the diameter of D job. And after that if we had a simple roughing or finishing insert at that time what we was doing. We simply write job diameter in geometry page. But here we have drill and drill have its own diameter. And that is why you have to first touch the drill to the diameter of the job. And then we have to go in geometry offset. And then in x we have to add job diameter and drill diameter and whatever value we get that value we have to write there. Means firstly does the drill to the diameter or the job. And then add job diameter and drill diameter. And that value we have to write in x. And after that press major key. So in roughing or drilling insert we simply write job diameter. But in drilling we have to add drill diameter in it as well. And z offset procedure is same means simply turns the tool to the face of the job. And then write z 0.0 in geometry offset and then press major. So I will show this procedure with the help of project so that you can understand it much better. So here I have taken for M drill. So firstly I will take this drill near our job. And after that, I will touch this drill to the diameter of the job.

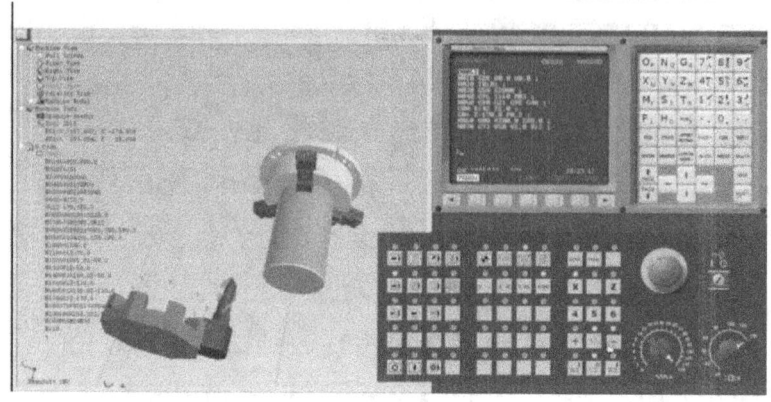

So before that, I will rotate spindle. Now I will touch D drill to the job diameter. Slowly. After that, you have to press offset key. And in offset, you have to go in geometry. And in that you have to select tool number that we have touch to the job diameter. So here my tool number is six. So in that I will select x. And here I will add job diameter plus drill diameter. So as drill diameter is 40 and we have taken job diameter at 150. So we will write here 190 in x. And after that we will press major. So in this way we have to take x offset for drill. And Z offset. You already know simply to the tool to the job face. Then write z 0.0 and then measure. So see here that procedure as well. Okay. So in this way, we have to give upset for drill. So this is our offset procedure. Now we will see our drill cycle. So here I have written one simple program in which I have included G70 for cycle. So as you can see G70 for ARM. So I have taken our value at 1.0 means tool will retract by one arm. In next block we have G70 for here. Ax will not come as we don't do drilling in X, we take good in Z and we will take curve in Z up to -80.0. And that is why I have written here z -80.0. After that p. So p is depth of cut in x as we are not going to take cut in x. So this p also not come here q is depth of cut in z. So here I have taken 2 a.m. cut. So we convert to M into microns. So that will come 2000. And after that feed 0.1. So now I will run this program. So see whether our G70 for cycle makes drill or not. So see, tool is not cutting material in one cut only. It is moving back and forth to cut that material means our G70 for cycle is working here. So this is all about final control. Now we will see this drill cycle in three months. Control as well. So for that, you have to create one new program. Here I will give name of our program as drill. So I have created one drill program. So in that I have to select one workpiece. So I will take one workpiece here. After that we will write first four five lines as per our standard format. As they are common for all the programs. After that, I will select Drill Tool here. So you can directly call that tool by writing that tool number. Or you can select it in this way. After that we will rapidly call our tool x 0.0 Z 2.0. And after that we have to write drilling cycle. So for that select drilling. And in that you have to select deep hole drilling means in this option you have to select deep hole drilling. After that you will get one form. Here. So you have to fill that form. So input will be complete. Blinn. You have to give x y blend as it is.

Don't disturb this spline. You have to keep x y plane only. After that we have our p means retraction. So I will give here to M m retraction as c means safety distance or safety clearance. So here I am giving one m safety distance. After that we have single position. So we have to keep it in single position. Here we have two options position pattern and single position. So select single position as we are going to take only one drill. After that we have cheap removal. And second is cheap braking. So in cheap braking tool will take cut and retract very less. And due to this small retract when g breaks into small pieces. And second method is cheap removal. Main tool will take cut and then comes out of the job.

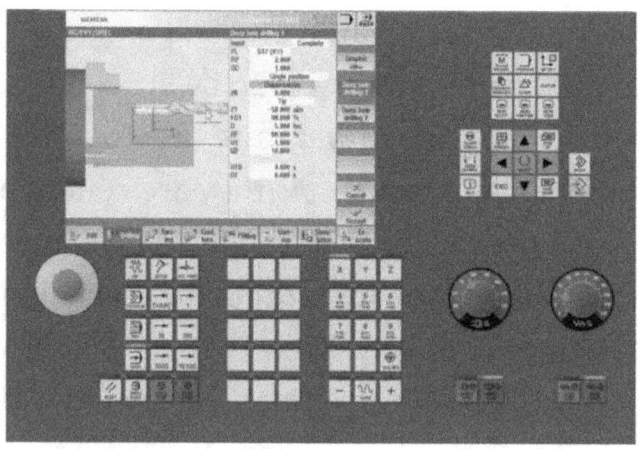

And due to which jibs that produce during cut will also comes out. And that's why it's name it cheap removal. So here we will select this cheap removal method. And this is much better method as this drilling works in ID and during I returning lots of it gets produced. And due to this hit there are chances of tool damage. But in this matter, as tool comes out of the job and do it to which coolant will reduce some heat. And this will avoid our tool damage. So that's why we will make our drill with cheap removal method. After that, Z0 means our starting point which is zero zero. After that, z one means up to how much distance. You have to do drilling. So that distance we have to write here. So we will do drilling up to -50. After that FD1 feed percentage for first in feed means how much feed you have to keep for first cut in drilling. So we will keep this feed as 90% after that. First, drilling depth means how much curd tool will take in its first cut. So we will write here to M and you have to take this in IMC means in increment means. Firstly tool will take two MB cut. And after that we have to write its feed. So we will take our first cut with feed 90% of our given feed. After that we want minimum depth in feed means after first cut how much depth of cut we have to take for next cuts. So here we have taken first depth of cut of two m, and after this what will be our depth of cut for next cuts that value we have to write here. So we will write here 1MM means firstly tool will take two MB cut. And after that tool will take one one MB cut. After that li distance. So we will keep this automatic. And this d tb d t d t s. These all are related to dwell time means how much seconds you want to hold your tool D s means how much time you want to hold your tool at starting position. D TB means after first cut how much seconds you want to hold that tool. And after last cut how much seconds you want to hold your tool. So that seconds you have to write here. So we will write 0.6 seconds. And after that press accept key. So this cycle will get added to our main program. After that we will take our tool out of this job up to Z 2.0. And after that M30 and off program. So in this way you have to make this drilling cycle in Siemens control. So now we will simulate it so that you will understand this cycle very well. So C tool is coming out of the job completely after cut. C means tool is cutting material according to our instructions. So this is our drilling cycle. So we see this drill cycle in Fanuc control as well as in Siemens control. Our next cycle is group cycle. So for growing we use G 75 code in this also. You will get two blocks. So in first block we have G 75 and R. So this R is retraction amount means after taking cut how much distance to L have to lift up. As we have seen in drilling the similar way we have this in grooving. And in our second block we have g 75 x means groove. Diameter means up to how much diameter you have to do grooving. After that Z last group position in Z axis means our groove. Last position we have to write here, be it depth of cut in x means how much depth of cut you have to take in x that value you have to write in p and in Q we have to write stepping in z axis means after taking cut in x, how much distance you have to move this tool in

Z. So that value we have to write here. And this p and q values are in microns. After that we have feed. So this feed you already know. And after that relieve amount at the end. So whether you have to give any relief amount at the end. So you have to write here zero zero as this is rarely used. So now we will find these values of our g 75 cycles. So G 75 and r. So this r is retraction amount. So here we have given 1.0 retraction amount means after taking cut tool we lift up by one m. Second block is G 75 x. So x means groove diameter or depth. So up to how much diameter. We have to make groove here. So here as you can see we want this groove up to 60 diameter. And that's why we will write here 60. Z means last position of our groove means let's suppose we are going to make this first groove and last position of our first groove distance. If we measure so that distance is 30 M. And that is why we write z -30. So here in grooving, you only have to call your tool above the groove position, and after that you have to apply G 75 cycle.

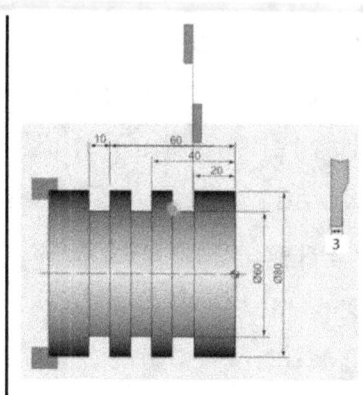

G75 R 1.0

G75 X 60.0 Z -30.0 P 3000
Q 2000 R 00 F 0.1

So tool will cut material automatically. After that, lift that tool and take that tool above this second group position. And after that again apply G 75 cycle. And there also tool will move up down up down to make that cut. So this G 75 cycle is used to make tool up down up down to take that cut. And that is why we write here Z -30.0 means here we have to make first groove and last position of our first groove is -30. After that we have p depth of cut in x. So this is in microns. So we will take three m depth of cut. And after that q stepping in z axis. So as we have seen after taking cut, if we move our tool by three M in Z. So there are chances that some material will remain. And that's why we always do stepping less than our groove thickness. So here our groove thickness is of three m. So we will do stepping in Z with two to m adds. If we move tool with three M then some material will remain in middle. And that is why we do stepping with two to m in Z direction. And this we have already seen. So we will write 2000 in Q as this value is in microns. After that our relief amount. So keep it zero zero. And after that feed 0.1. So now we have to call our grooving tool. And as you know this phase of our tool 800 means this with is touch to the jar phase. And then we take geometry of Z. And that is why if we call our tool at z -20.0, then that tool will stand here. But we don't want to make groove here. We want groove here. And that's why we rapidly call our tool above 80. Diameter means in x above 80. And in there we will call it at -23. So our tool will stand here. And after that we have to write G 75 cycle. So tool will do up down up down movement to make that groove up to Z -30.0. So this we have already seen in our example topic. And here also we have covered everything. So here I will write a small group program in final control. And we will see whether G 75 makes groove or not. So see here I have written one small group program. And as you can see I have called tool at x 100.0 and Z -23.0. And after that I have written these two G 75 blocks. So I have written these two blocks for our first group. And we have to make three group. So that's why I have left that tool after first group up to 85.0. After that I move that tool in Z up to -43.0. And after that, I have written these two blocks of Z 75 cycle. And I have done this same procedure for our third group as well. Means after making second group I have lifted that tool, move it in Z direction. So up to how much distance we have to move that tool in Z. So -63. And after that you have to write G 75 cycle. So it means whenever you want to make groove just tool about that group position and then simply apply G 75 cycle means right that two blocks of G 75 cycle. So tool will automatically make that groove. So now let's run this program. So firstly tool stands at Z -23. And now it will make groove. So it made that groove. And now we have to make second groove. So we will travel it in Z up to -43.0. And then we will apply G 75 cycle. So tool will make that second group as well. After that we have to take that tool up to Z -63.0. And then write G 75 cycle. So in this way we made this three grooves which is 75 cycle. So now we will do measurement of our groove so that there will not be any doubt in groove. So you can see in Z distance here first position is at -20. So we nearly reach up to -20. After that this group is of ten. So 20 plus ten is equal to 30. So this distance must be 30 means -30. After that -40 after that -50. After that -60. And in last -70. And if you can see groove depth in x which is 30. So we reach nearly x 30 means its diameter is 60. So you can see in diameter which is 60 as we wanted. So our dimensions are accurate. And our G 75 cycle run properly. So this is our grooving cycle. So we saw this grooving cycle Inferno control. Now we will see this grooving cycle in Siemens control. So for that firstly go in program Manager and then create new program. And you can give any name to your program. So here I will intentionally give some different name drill ABC. After that we have to add our workpiece. So take atm workpiece. After that right. First for five lines as per our standard format. After that, we will select our grooving insert. So from here we will select it. Or you can directly call it. After that, we will rapidly call our tool. Now we have to make groove. So for that we will go in turning option.

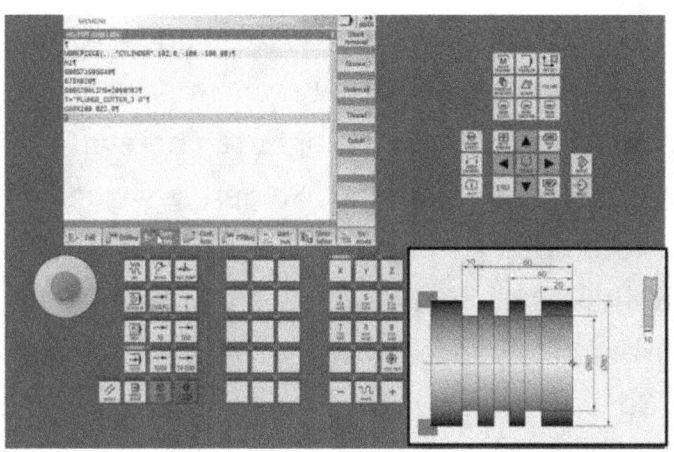

And in that turning option you will get groove. So you have to press it. And now we will make simple group here. So for that we will keep first position. And second option is like this. If you want this type of groove then select second option. So you have to choose this according to your requirement. But we want simple groove. So select first option. And in that you have to enter here values. So SC fs you already know this. Then we will make roughing as well as finishing both with this same tool. So x zero and z zero means we have to give dimensions at this point. So here we will start doing groove from last means. Firstly we will make last group, then middle and then first. In this way we will make this groove. Normally we make groove from starting. But here we will do something new. We will make these groove from last. So that round part will come here. And we have here x 80. And if we measure Z then it will be 60 means - 60.0. So we will put these values in x zero and z zero. After that we have groove width. Be one means groove width. So groove width is ten. So right ten here after that T1 means groove depth diameter. So here we want this groove up to diameter 60. So we will write this 60 in T1. After that we have depth of cut. So we will take 0.5 depth of cut. After that allowances in x direction. Allowances in z direction. So write these values and end means number of grooves means how much grooves you want to make. So here we are going to make three groove. So we will write n is equal to three. After that it is asking DP means distance between this point to this point. So how much distance it is. So if you can see here. So from this point to this point distance here is 20 M. So if you calculate you will get this distance as 20. And after that press exit. So here we have given all the information. After that we will lift our tool up to x 100. After that we will do tool homing and then M30 end of program. So in this way you have to write this groove cycle in Siemens control. So let's see whether it really works or not. So see firstly it is making last groove. After that middle one, as we are making this groove from last to first. And then first groove. So here our all three grooves are ready. That means our cycle is accurate. So in this way, we have seen this groove cycle in final control as well as in Siemens control. So this is our groove cycle. Now our next cycle is three trade cycle. And for this trade cycle we use G 76 chord. So in G 76 cycle also you will have two blocks. So in our first block we have G £0.76. And in this p we have three values. So in first position means in P1 we have number of finishing passes means after completion of thread formation we have to take two three finishing passes. So that if there is any power or material remain so that will get us removed with this finishing passes and you will get clean thread after that P2 Java amount. So if you want any chamfer in your thread then you can write that value here. Normally we take this value at zero. So you also write here zero. As this chamfer in thread is made in special condition. So write zero zero here. And third is thread angle means you have to write angle of your thread here. Normally we take 60 degree thread angle. After that we have q depth of cut for roughing means how much depth of cut you have to take for roughing process. And this value is in microns and R is depth of cut for finishing means how much cut you have to take for finishing. And this value is in m m. So you have to remember which value you have to give in micron S and which value is in M. Our next block is g 76 x. So here x is miner diameter of thread. And that we have already seen. So that diameter we have to write here. Z means thread length means up to which length. You have to make this thread. That value will come here after that p thread depth. So we have seen how to calculate this thread depth means when we calculated miner diameter of thread. At that time, we have calculated this thread depth and then we calculated miner diameter. So that thread will come here. And this thread that value is in microns and q is depth of first cut. Means how much cut you have to take for first time that you have to write here in microns. After that f means feed. But here we have to write thread page. So thread page we have to write here in feed. And these are its taper value means if you want to make this thread in taper then we have to give that taper value here in R. So firstly we will make simple thread and then we will make taper thread. So now let's calculate all the values from this cycle. So in our first block we have g £0.76 and in p we have three values. First is number of passes. So here I am giving two finishing passes. After that chamfer amount. So as I told you keep this value at zero zero as this is for special condition. So you have to keep this value at zero zero and then thread angle. So here we have taken 60 degree thread angle. But you have to check this thread angle value in your drawing. This thread angle value is there in drawing after that queue. So q is depth of good for roughing and this value is in microns. So here I have taken depth of cut adds 100 microns. And after that depth of cut for finishing. So here we have taken 0.0 to m depth of cut as this value is in m m. In our next block we have g 76 x. So x means miner diameter. So this miner diameter formula we have already seen. So firstly we are going to do this first thread M 20. So miner diameter for that thread will come 18.17 m m. So we will write this value in x and up to which distance we have to make this thread. So up to -20. So we will write z -20.0. Be is thread depth. And this is in micron. So we will write this value in micron here 9150 after that q. So q is depth of first cut. So how much care do we have to take for first cut. So we will write 200 microns means our first cut will be of 200 microns. After that feed main speech. So our pitch is 1.5. So we will write here 1.5. And we are going to max thread thread. We are not going to add any taper in it. So we will write here in R at zero zero. So in this way you have to write this thread cycle. So now we will run this cycle in final control. So here I have written simple program in which I have included G 76 cycle means two blocks of G 76 cycle. So you have to see whether with this cycle our thread forms or not, all values I have written here that we have calculated. So now let's run it. So here tool made that thread. So our G 76 cycle is also working. So here I have given some important notes about thread insert.

Imp notes for thread insert

If we want right hand thread(mostly used)

1. Use Right hand insert(ERH written on insert)
2. Tool holder must be right handed
3. Spindle rotation must be anticlockwise direction.
(we always fit tool oppositly in turret)

So if you want to make right hand thread that mostly yields means mostly this type of thread is used right hand thread. So if you want to make right hand thread at that time, use right hand insert. Only e r edge will be written on that insert. You have to take that insert only. And our tool holder must be right handed as well. So we have seen this right handed left handed tool holder. So you have to select right handed. Tool holder means the right handed tool holder with right hand insert. If you want to make right hand thread and spindle rotation must be anticlockwise. So you have to look these three conditions whenever you want to make right hand thread. So you are insert must be right handed. Insert and tool holder also must be right handed and spindle must rotate in anticlockwise direction. And this tool we are going to feed into it. So that tool must fit oppositely in that turret. So if you have not this thread tool holder is always created in reverse direction. So in the same way you have to feed this tool holder. And then only we will get this right hand thread. If you feed this tool without reversing it, in that case our tool body will touch this spindle and do it to which our tool may get damage. Because if we feed this tool without reversing, it means if we normally feed this tool as we do for roughing or finishing. So in that case, our insert will left behind and body will move ahead. As you can see here, our insert is behind and body is ahead. And that is why we always reverse this tool holder. So that our insert will be ahead and tool body will left behind so that there will not be any accident. Okay. So this tool holder is always fitted in reverse direction. So this is all about some important information about thread insert and thread tool holder. Now we will see it threading. We have already done already threading calculations. Now we will see some calculations for side threading. So there are some changes in formula for finding x means inner diameter and why we have that change that also you will know. So minor diameter means up to which diameter you want this threading. So what is our major diameter here. So suppose this diameter is 20. So as you can see our minor diameter is greater than major diameter as we are doing threading in I-80. So this is our major diameter means diameter on which we don't have threading. And minor diameter eight diameter that forms after threading. So after threading, diameter increases in ID. And that's why minor diameter is always greater than major diameter in ID. So that formula to find minor diameter is major diameter plus two into depth for. Already we had major diameter minus two into depth. And for ID we have major diameter plus two into depth. So here major diameter is 20. And add this two into 0.915. So this is 0.915 is depth. So how to calculate this depth that we have already learned. So if you do this calculation then that will come 21.80 3MM. So for already we have here minus and for 80 we have here plus because inside our minor diameter increases after threading and in already our minor diameter decreases after threading. So this is the only difference in calculating x for id and o.d. So now we will see how to make taper thread. So let's suppose our starting diameter is 16 m and our second side diameter up to which we want our taper thread which is 18.17. So we want this taper thread like this. So for that we have to find our value means in thread cycle we have. And so we have to give some taper value there to get taper thread. So how to calculate this r value that we are going to see here. So R is equal to starting diameter minus diameter up to which we want this thread divided by two. So this is the formula to find r value. So if you do this calculation so we will get -1.085. So our thread cycle will be same. Only you have to write r value adds -1.085 so that you will get taper thread. So now I will show you in final control that how this taper thread forms. So as you can see here I have written r value adds -1.085. So that tool will make taper thread remaining everything its same. So now we will run it. As it is not properly visible. So we will see this with measurement view so that we will confirm that its taper thread. So see our thread is in taper here we have large diameter and this diameter is decrease in continuously means its taper thread. So you have to write our value in our G70 six cycle to get taper thread. And how to calculate this R value that also we have learned. So in this way we have to make this taper thread. So this cycle we have seen in value control means this thread cycle we run in value control. Now we will see this thread cycle in Siemens control means we will see how to make threading cycle in Siemens control. So for that first press program manager key and create here new program and give name to that program. So here I am writing name as thread. After that we have to add workpiece. Here. So add workpiece take diameter at 20. And we are going to do threading on 20 diameter. After that we will write starting for five lines as per our standard format. These lines are common in all the programs. After that, select thread tool or you can directly call it. So here I will select that tool. After that write D12 cancel preceding upsets of that tool. So these we already know after that rapidly call our tool. And now we have to do threading. So for that select turning option. And in that option you have to select thread. And after that you will see a form. So you have to fill that form. So input will be complete in table. We have isometric and more unit options but you have to select none here. After that we have pitch. So we will take thread pitch as 1.5. And G means if you want. This type of thread means firstly we have pitch, then pitch plus g value means this thread width will increase, after that it will increase more and more. So if you want this type of thread then you have to give value here in G. But we don't want this type of thread. We want thread with same page. And that's why we will write in G as zero zero. After that machining. So here we will do roughing plus finishing with this same tool together. And that is why we have selected this option after that linear. So here they have given two methods to take cut. First is linear and second is digressive. So in digressive we have this type of cutting means in starting it will take maximum cut, then it will take less cut, then again less cut. So in this way that tool will take cut. So we don't want to take these type of cuts. And that's why we will select linear option means it will take equal cuts in every path. After that we are going to make external thread. Internal thread means it threading. So we are going to make already threaded. So external thread. So now we have to write x zero and z zero means as you can see this point. So you have to write coordinates at this point in x zero and z zero. So our x zero is 20 and z zero is 0.0. As we have taken raw material at 20 diameter. After that we have z one means up to which length. You have to make this thread. So we will make this thread up to -50. After that LW means this distance means after taking cut how far we want to stand our tool from your face. So we write here 2MM means we will stand our tool two m away from the job. Face after taking cut. After that LR means in last tool will take cut in angle. So if you want this angle in last then you have to give distance from which tool start making angle. If you write zero zero then tool will make straight cut. It will not lift in angle at the end. So if you want this angle in and then give distance here. Otherwise write zero zero for straight cut after that h one depth of thread 0.915. Here machine automatically calculated 0.920. So we will keep this value as it is after that. Infinite slope angle means thread angle. So here you have to write half angle of your thread. Now suppose we are going to make 60 degree thread. So its half its 30 degree. After that D means depth of cut. So we will take 0.3 depth of cut after that this distance. So we don't want this type of any distance.

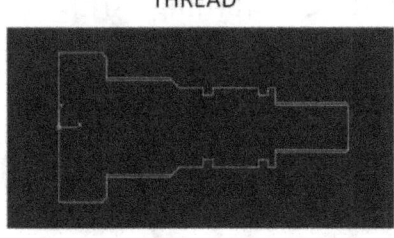

THREAD

N3;
G28 U0.0 W0.0;
T0404;
G00 X12.0 Z2.0;
G76 P02 00 60 Q 100 R0.02;
G76 X8.17 Z-15.0 P9150 Q200 F1.5 R00;
G00 X22.0;
G00 Z-35.0;
G76 P02 00 60 Q 100 R0.02;
G76 X17.865 Z-50.0 P1067.5 Q200 F1.75 R00;
G00 X25.0;

P=0.61*PITCH
=0.61*1.5
=0.915

P=0.61*PITCH
=0.61*1.75
=1.0675

X=10-(2*0.915)
= 8.17

X=20-(2*1.0675)
= 17.865

So we will write here zero zero. After that how much finishing passes you want to take. So here we will take three finishing passes. After that. This distance means your tool will take cut. And after that up to how much distance you want to lift. It means after threading how much distance you are going to lift that tool. So that value we have to write here. So we will lift that tool by two m. Now I hope you have learned how to fill this form. You only have to see what parameter it is indicating exactly. And according to that you have to insert values don't change everything. You only have to change important parameters. Remaining everything will be default. Our machine calculated values. So you only have to input important data in that form. So until now we have filled many forms. And from this you might have get an idea that how to fill this form. So from now you can fill any form in Siemens control. And after that press accept. So we will get this cycle here. After that we have to lift our threading tool. And then we will do tool homing. And after that and that end of program. So this is our threading cycle. So now let's simulate it. So tool is making thread. That means our cycle is correct. And according to that information tool made that thread. So now we will see how to make taper thread in Siemens. So for that also create new program. You name to that program. After that we will add here workpiece. Now you know all these things. After that right. Starting for five lines as per our standard format. After that, take. Threading. Insert. Then rapidly call it. So for taper thread select turning option. And in that select thread and in thread select taper thread. For simple thread select thread longitudinal. But now we will make taper thread. So input will be complete. Now you know all these pages will be 1.5. And we do not want this type of thread. So we will write here zero zero. After that we will do roughing plus finishing together with this same tool. We will take linear means. It will take equal cuts in every path means equal depth of cut. Then external thread. Then x zero and z zero. So let me take x zero at ten and z zero at zero zero means we will take our workpiece of ten m m after that. Thread flow angle means in which angle. You want this thread that you have to write here after that z one. So we will take here -100 means I want this thread up to length -100. So Z1I have written -100 after that LW. So LW means this distance means after taking cut how far you want to stand that tool from your face. So that distance you have to write here. So I will take that distance as 1.0. After that LR means this distance means in last tool will make thread in angle. So if you want this then you have to give distance here. And from that distance tool will start making angle. If you give here zero zero then tool will not make any angle. So last time we write zero zero. So now we will write here two. And we will see how tool makes that thread. After that H1 depth of cut or depth of thread 0.915. Here machine automatically calculated 0.920. So we will keep this 0.920. After that infer slope angle. So you have to write half value of your thread angle. So we are going to make 60 degree thread. So half of 60 degree is 30 degree. So we will write here 30 degree after that D one depth of cut. So we will take 0.3 depth of cut. So I don't want this distance. So I will write here zero zero. Then finishing passes. So I will take three finishing passes after that. Up to how much distance you want to lift that tool. Means after threading up to how much distance you are going to leave that tool. So that value we have to write here. After that, whether you are going to make multiple thread. So select here. No we will make single thread after that starting angle offset. So we don't want any offset. And that is why we will write here zero zero. And after that accept. And after that you have to lift that tool. And then do tool homing. And in last right. And 30. So now we will simulated and we will see whether the cycle makes simple thread or taper thread. So see tool make that thread in taper here. Tool directly start making threads as we are doing only threading. But in practical we don't do this. Firstly we will do roughing and then finishing. And then we will make this thread. Here you only have to see whether this cycle makes taper thread or not. So as per our cutting parameters we get our taper thread. So here we have learned our threading cycle. So we run this cycle in value control as well as increments control. So now we will see one example in which there will be all operations. With this problem. We have already seen in example topic. So we will take this same example and we will make program for it with cycles only means we will only use cycles to make program for this diagram. So now let us start it. So firstly we have to make this type of job from our raw material. And for that we will write G70 one cycle. So some starting lines are same as per our standard format like to call coolant on. So this will be same after that. We have G70 one cycle. So if you remember our G70 one cycle. So in that cycle our first block was you and our means G70 one you and our. So this you eat depth of cut in X. So here I have taken depth of cut in x adds one m and our lift after cut means how much distance you want to lift that tool after taking cut. So here we lifted our tool by one m and in next block we have G70 one speed. So finishing program that we are going to write. So starting block of that finishing program will come here. And in queue we have last block number of that finishing program. So this last block will come in queue. After that you. So you is finishing allowance in X. So we will take allowance at 0.5 m. W is finishing allowance in Z. So here we have taken zero point 3MM and fed. So here I have given 0.3 feed. And after this we have to write our finishing program here. Now you try to make this finishing program by your own. So firstly calculate its coordinates. So we have already learned how to calculate coordinates in our example topic. So you can easily find coordinates for this job. So read you are drawing carefully and find coordinates for it. And then try to make finishing program for it. You only have to write finishing program here. I have already written this finishing program here for you, but you have to try this by your own. Now we are in last topic of our programing syllabus. So you must be able to do this. So write this finishing program here. And after that we will write g70 P and Q. P means starting block of our finishing program. And Q means and block and then tool change. If you want to change your tool for finishing then you can change it. So in this way we write this G 71 and G 70 cycle. So you try to make this finishing program. If you have any problem then you can refer this program. I have written here for you. So after that we have to lift our tool. So this is our G 71 and G 70 cycle. Now we have to make group in it. So for grooving cycle we have got G 75 so right. And two and then two tool homing. Then call grooving tool. And then we will rapidly call that tool in this position. So this position is x 18.0 and z -18.5. So you can see this in dimension means we stand our grooving tool above this groove. And now we will write our grooving cycle. So G 75 R. So these are its retraction amount. Means how much distance you want to lift that tool after taking cut. And our second block is G 75 x. So this x is grooving diameter. So our dis grooving diameter is 13. So we will write here 13. And z means last group position. So you have to write this distance. So this distance is 18.5. So we will write here z -18.5. After that p. So this p is depth of cut in x. So here we will take one more depth of cut. So if we convert this in microns then that will be 1000. Do it stepping in Z.

So here we will make this two m group. And we will take two m groove. Insert. And that is why there is no need to do stepping in Z. And that is why we will write here in Q at zero zero after that relief amount. So you have to take this value at zero zero. And after that feed. So this feed we will take 0.1. After that we will lift our tool. And then we will move it to Z. -30 means we stand our tool about this groove. And again we will use grooving cycle here. So G 70 5RR is retraction amount 1.0. In second block we have G 75 x. Our groove diameter is 30. Z is -30. Means last group position eight -30. Then p depth of cut. So we will take one more depth of cut. So we convert one m into microns. So we will write here 1000 then q. So we will not do any stepping as grooving width and grooving insert is of same dimension means we will make groove of two M and we will use groove insert of two m thickness. And that is why there is no need to do stepping in Z. After that our relief amount zero zero. And then feed 0.1. So that tool will make this groove as well. After that leave that grooving tool. So in this way we will make this groove. After that we will make thread. So right entry then we will do tool homing. After that we will call our threading tool. Then we will rapidly call it. And after that we will write G 76 cycle. So in G 76 cycle we have p and in p we have three parameters. So first parameter is number of finishing passes. So we will take two finishing passes. Here. After that second parameter is chamfer amount. So we will not make any chamfer. And that is why we will write here zero zero. And after that trade angle. So we will take thread angle at 60 degree. So this is p. After that we have q. So q is depth of curd for roughing. So we will take depth of cut for roughing as 100 micron. As this q value is always in microns and R is depth of cut for finishing. So we will take 0.02 m depth of cut for finishing. So this value is in m, m and in our second block we have g 76 x. So what is the miner diameter. So if you calculate miner diameter for this first thread. So that will be 8.17 M. So right here 8.17. After that Z means up to which length you want this thread. So we want up to -15. So we will write here z -15. After that p means depth of cut. So this is in micron. So convert 0.915 into micron. And right here 9150. After that q depth of first cut. So we will take our first cut of 200 microns as we have to write this value in microns after that feed. So here we write pinch of thread. So as our thread page is 1.5. So we will write here 1.5. And after that paper amount. So we will write here zero zero. As we are going to make simple straight thread here. And that is why we will take this taper amount adds zero zero. So here we have written all the parameters for our first thread. Now we will leave that tool up to x 22.0. Then move it in Z up to -35 and then again write threading cycle. So in first block we have g £0.76, and in p we will write 0200 and 60. Now you know these parameters. First parameter is finishing process. So we will take two finishing passes. This is chamfer amount. So write zero zero and then threading angle. So we will take

60 degree thread angle then q. So this q is depth of cut for roughing. So we will take 100 micron cut. And for finishing we will take cut of 0.02 m after that. In second block we have G 76 x. So this x is miner diameter. So if you do this calculation then you will get 17.865. After that z means up to which then you want this thread. So if you see this distance then it is -50. After that we have to write depth of cut in microns. So write this value here. And after that we have q depth of first cut. So we will take first cut of 200 microns. And after that which 1.75. And then taper amount zero zero. As we don't want any taper in thread. So two L will make this second thread as well after that lifted. Now we will do drilling. So and for then tool homing then tool call. After that rapidly call it up to Z 2.0. And after that we will write G70 for cycle. So in first block we have G 74 and R. So this R is retraction amount. And in our second block we have G 70 400. As we are not going to take any cut in X we will take it in Z. And that's why x will be zero zero. And z means up to which distance you want drill. So we will take total cut. And this distance is 60. So we will take cut up to -62 for proper cutting of material. After that p depth of cut in x. So this will be zero zero as we are not going to take any cut in x. And how much depth of cut you want to take in Z. So we will take 20 micron depth of cut as we have to give this value in microns after that feed. So we will take 0.1 feed. And after that we will write G00Z 2.0 means we will take that tool out of the job. So in this way we will make our first

drain and now we will make six m drain. So write N5 then do tool homing. And after that call 6MM drain. Then rapidly call it up to Z 2.0. And after that we will write G 74 cycle. So in first block we have G 74 are retraction amount. So we will take 1MM in next block. We have g 74 x as we will not take any cut in x. So we will write here 00Z will be -20 as we are going to do. Drilling up to -20 only means six m drill. We take up to -20 after that p depth of cut in x, as we are not going to take cut in x. So this value will be zero zero. Then q depth of cut in z. So here we will take 20 micron cut. So right here 20. And after that feed. So we will give 0.1 feed. After that take that tool out of the job up to Z 2.0. And after that we will do tool homing and M05 and zero nine and N 30. So our total program is ended here. We make program for this job only by using cycles. So we make total program with the help of cycles. So now we will run this program that we have written here in Fanuc control so that you can compare your cycle with actual tool movement. So see this project here I have written all program that we have recently made. And I have included all cycles with everything. So you just check it. Every cycle is there in this program. So now we will run this program in single block so that you can understand much better. So firstly, tool will take cut according to G 71 cycle means our stock removal roughing cycle. After that, finishing means G 70 cycle. After that groove. So see, we are doing this in single block mode. After that thread. In both the places. After that drill, we will do drilling two times. First, leave it for M and second with 6MM. And this 6MM drill. Okay. So our all cycles are working properly. And we got our job exactly as per our requirement. So now we will write the same program in Siemens control. And we will run this program in Siemens control as well. So here I have written all the program means I have created all these cycles. Here you can check it. So now we will simulate it. So firstly tool will take cut according to stock removal cycle means. Firstly it will do roughing. After that finishing off stock removal cycle. And after that threading. After that grooving. And after that. Drilling. First for m m drill and second 6MM drill. So tool will make these two drills. Firstly, it is making for m m drill. And now 6MM okay. So our program runs properly in Fanuc control as well as in Siemens control. So here we have completed our cycle topic. We learn all cycles and run that cycles in FANUC control as well as in Siemens control. So we have learned everything theoretically as well as practically. You have to practice these cycles as much. You practice these cycles, you will feel much easy. So here, as you can see, we have made programs in two methods. First, we write programs with G00 and g zero and code. And secondly we use cycles. So now you will ask which method we have to use means either we use G00 or G01 to write program, or we have to use cycles. Or when we have to use these two methods. So there is no such kind of restrictions here. You can use any method anywhere, or you can use both. The methods together to write program means some program you write with G00 and G01 code, and some part of that program you write with cycles. This also works. So what happens if your job each two week means it's cycle time? It's large means nearly. Suppose half an hour or one hour or your material. It's so hard so that more number of inserts gets we are out. So in that case, you have to make this program with G00 and G01 code. So let's suppose you write program which have large cycle time. And you write that program by using cycles. So in cycle we only write finishing program. And about that we write two lines of that cycle. So let's suppose your program is running and it runs properly up to first 20 minutes. But suddenly your insert fails. So for changing insert you have to stop machine and then only you will able to change insert and after changing insert program will start from starting or it will start from that cycle. If you have written program with cycle. So in this case our 20 minute program that already done will start again here in cycles. We don't have any other solution to start program from where it has stopped means from 20 minutes. So this is not possible in cycles. So this is one disadvantage of using cycle. And that's why if you have not noticed, large cycle time programs are written with G00 and G01 codes. Add in this. If insert fails then you can start that program from where it has. Stop means you have to select block number from where you have to start that program, and then just press Cycle start. So in this way you can start program from anywhere in G00 and G01 code. But in cycles your program will start from its starting. So you G00 and G01 code for writing large cycle time program. And you can use cycles in small cycle time jobs, as there will not be that much loss of time if you start that program from starting. So you can use cycles in small cycle time programs, or if you have to develop only one job. So in that case, also you can use cycles. But if you have large cycle time job then make program for it with G00 and G01 codes. So here we have completed our programing part as well. After this topic we will see some small concepts like copy paste and all that we will cover in our next topic. So here we have completed our cycle topic.

MORE INFORMATION

Our next topic is more information. Until now we have seen our programing syllabus. So in this topic we will see some extra knowledge that may help you in future. So first is Big Ed. It means background ed. It means what happens if your machine is on means any program is running on machine. And if you want to create any new program in that running condition, in that case you have to use big Edit. And this will reduce your time as you board to work is going on together means program is running on your machine as well as you are creating new program or editing program and data, which you will save a lot of time. So let's see how to do this. So firstly we will see how to create new program. So for that you have to select program key. And in that select ability after that big edit. And in that select edit. And there you have to write program name that you are going to create. And after that insert. And then start writing program. And after that big end. So your program will be ready here.

Create new program

Program ⟹ OPRT ⟹ BG Edit ⟹ Edit ⟹ enter number(e.g O0050) ⟹ Insert ⟹ Start writing program ⟹ BG End

Editing existing program

Program ⟹ OPRT ⟹ BG Edit ⟹ Edit ⟹ enter number(e.g O0050) ⟹ OSRH ⟹ Rewind ⟹ Start editing program ⟹ BG End

You can also do edit in your program when your machine is in running condition. So any changes you do in edit get saved in your coffee machine. So we have seen how to create new program. And now we will see how to do editing in program when machine is in running condition. So for that select program key then operate that edit edit then write program name in which you want to do editing. And then o s r h. After that press rewind option. And after that start editing your program and then press BG and so in this way you can do editing. So now we will see project in which we will create new program. And also we will do editing in it. So firstly I will run this machine means I will rotate spindle. So now machine is in running condition. So press program key then BRT after that BG edit. Then again edit. And now we have to write name of program that we are going to create. So we will give name of our program as 201. And now start writing your program here. So here I will write one small program. I'm going to create very short program of two three lines. So our program is ready. After that you have to select add option. So our program is created and save in machine. So let's check whether our 201 program is really saved in machine or not. So for that we will search our program. So select program and then write O0201.

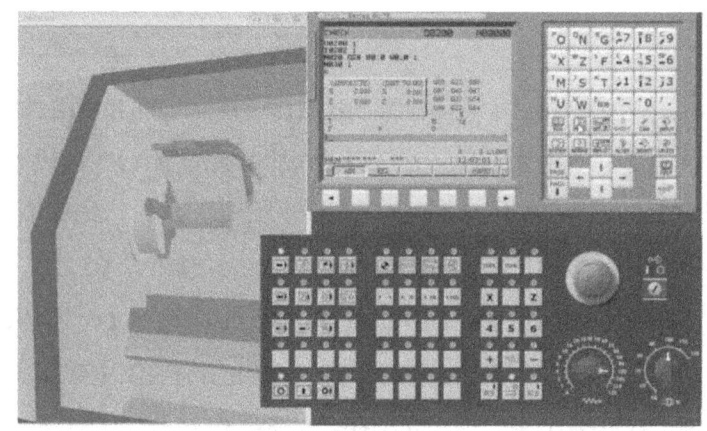

Now you know how to search program and then press this key. So our program open means it is saved in machine means our machine was in running condition. And with that running condition we created new program. So now we will do some editing. So I will rotate spindle again so that our machine will be in running condition after that press program, then appear then edit then edit. And now call your program in which you want to do editing. So I will write 201 in which I want to do editing. After that o s r h and then rewind. So our program open here. So now do editing in it. So here I will change to number 2T0202. So we change our tool and BG end. So if we search this program to see that change to zero and program. So see our changes are saved in program. So we created new program as well as we do editing in it while our machine was in running condition. So in this way you can create new program. And also do editing with the help of edit. Now we will see how to do cut, copy and paste in this simulator we don't have copy paste option. Then also I will tell you how to do this copy paste in CNC machine.

CUT COPY PASTE

So firstly turn on edit mode. After that turn on lock key. And now we will open program in which we have to do copy and paste. So now you have to see where is our copy option here. So nearly here you will get this copy option. So now select line that you want to copy. So suppose I want to copy 30 number block line. So I will move cursor on it. And then I will press copy option. And now we have to paste it. So open that program in which you want to paste this line. So select auto mode. So now you know how to open program. So open that program. And in that program move cursor where you want to paste that line. So much cursor. There. And here you will get paste option. So simply press paste so that that line will be pasted here. So in this way you have to do copy and paste here. We don't have these copy paste options in this simulator. But you will get these options in your CNC machine. So this copy paste is very simple. Just move cursor on line which you want to copy and then press copy. And wherever you want to paste that line, move cursor on it. And then simply press paste. So that line will paste it there. So this is very simple. So you try this on your CNC machine. It's just like copy paste in our mobile.

CUT COPY PASTE

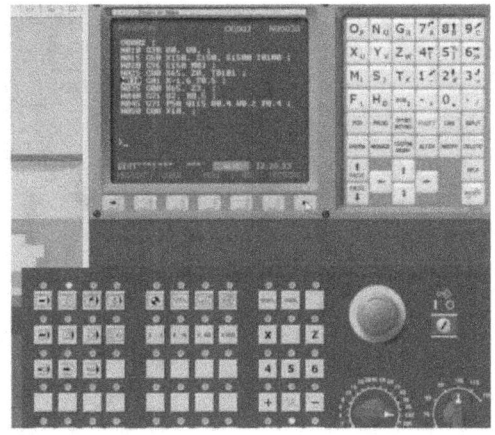

So you can easily do this. So now we will see how to take program from memory card to CNC memory. So firstly you have to insert card means memory card or pen drive. So in CNC machine there is a slot near display to connect this. So with that slot you can connect memory card or pen drive to CNC machine. After that you have to turn on the key means our lucky. So turn on that key and after that press MDI. Then you have to press offset two times. If you press offset key two times, then this window will open, and in that window you will get input channel option. So take your cursor there.

MEMORY CARD

Memory card to cnc memory

Insert card ➡ key on ➡ MDI ➡ offset(2times) ➡ i/p channel
(4-memory card)(17-pen drive) ➡ input

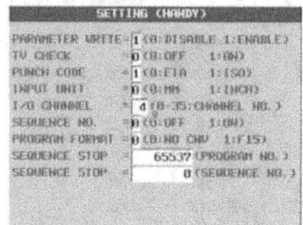

And here we write for for our memory card. And we write here 17 for pen drive. So if you have memory card then write for here. And if you have pendrive then write 17 here. And after that press input key. So with this correction machine will understand that it have to take information from memory card or from pen drive. So after that if you want to take program from memory card to CNC memory then let's see how to do this. So for that firstly press edit key. After that you have to press program button then DVR or folder. Whichever option your machine have that option you have to press after that press opacity button, then press device option and then memory card. And in that memory card, select whichever program you want in your CNC machine means just take cursor on it and then press F input option as we want that program. That is why we will press F input option. After that execute, then press device, and then when you want that program. So we want that program in our CNC memory. So select CNC memo option. So in this way you can take program from memory card to CNC machine. And if you want to send any program from CNC memory to memory card at that time, you have to follow this below procedure. So firstly press edit key then program, then D r or folder, then appear to then device after that memory card and in that whichever program you want to send that program, you have to select here and then press F output. That means that program you want to send from machine to memory card. That's why we press F output key and then execute. Then press device and then press C and z memo key. So in this way you can send program from CNC memory to memory card. So this is the procedure to take or send any program from CNC machine to memory card. So you try this procedure on your CNC machine. If you have any changes in option means if your CNC machine don't have option from this procedure at that time, instead of that option, your machine will have some different option as we have here. If your machine have D option, then select it. Otherwise your machine will have different option. So these changes differ from machine to machine. But this is this standard procedure to take program from memory card to CNC machine and from CNC machine to memory card. So this is its method. After that we will see part count means number of parts that we do. Machining means during operating you have to give report of your production. At that time, you have to give this part number. So before starting your production work you have to make part count at zero. And then you have to start your work. So let's see how to make part count at zero. So for that press or button. After that p t s p r e. So that part count will get selected. After that you have to press execute. So whatever number was present there will become zero. So in this way you have to make zero. Here simply press op parity. Then p d s p r e and then execute to make part count at zero. And if you want some other number here. So for that you have to press offset two times. And when you press offset two times. Then this window will open. And in this window you will get option here. Part count. So write any number here you want and then simply press input. So that number will display in front of part count. So in this way you can display any number in front of part count. So here we have learned every detail thing about CNC machine. So here we have completed our total programing course. We have covered everything from basic level to advanced level. So after this project we have given some guidelines to you about how to move further in your career. So watch that project as well and practice this. So if you have any doubt then you can contact us. We will try to reply you as soon as possible. So here we have completed our complete course.

CAREER GUIDANCE 3

So how are you all? I hope you have spent your three months in setting. So now you are seeing this set of means you can do setting perfectly as well as speedily. Now you can ask to increase your salary as you have knowledge. Now means you can do CNC operating as well as CNC setting. So now you can tell your boss that now I'm able to do setting as well as operating. So kindly increase my salary. But if by some reason your company is not able to increase your salary, then you have to try in another company and you will get this job very easily as you have knowledge now and you can do setting perfectly as well at speedily so you can get CNC set a job easily and your salary will also gets increase. Company not let their employees to go, they increase their salary so you can ask about your payment to your boss. But you don't have to think that much about your salary right now. As our thinking. It's so big and we want to earn more money than this. And that's why don't think about money only as we want more knowledge right now. So now you are seeing this etc. means you have spend your three months and then you have watch our programing topics. So now you have to spend next six months in CNC programing as this is programing. So you will require did six months more. You spend time in programing the more easily and fastly you can make programs. So what you have to do in this six month. So firstly watch programing projects again and again. Whatever knowledge you will get in these projects that you have to apply on your CNC machine. So this is common means you have already done. This means you have already done this every time. After that, what do you have to do? More? So whatever drawing you see in your company, you have to take that drawing in your mobile or take a photo of that drawing. And after coming back to home, you have to try to make program for that drawing. So you have to do this work for next six months more. You do practice of making programs. The more expert you will become. You just take photo of drawing, come home and try to make program for it. You only have to do this work for next six months. Watch our projects again and again and whichever drawing you see, take photo of that drawing and try to make program for it so more. You write programs more speedily and easily. You will able to write programs.

WHAT TO DO IN THIS 6 MONTHS

1. WATCH VIDEOS AGAIN AND AGAIN.

2.

So when you are able to make program for any job or drawing easily, then you have to do one thing you have to take for raw materials, you can get this raw material very easily in market. So take for raw material of different diameter and different length. Don't take big raw material. Take small medium size raw material so that you will get these raw materials at very cheap price. And in that raw material you have to keep two raw materials separately and next to raw materials separately. So in first two raw materials you have to check diameter and length. And according to that you have to make drawing for it. And in that drawing try to add as much operations as you can. And same way you have to do this for next two raw materials means by looking at diameter and length. Make trying for it and try to add all the operations in it. And after that you have to make program for first two raw materials by using G0 zero and G0 one code. Don't use any cycle in it, only use g zero, zero and G01 code to make that program. And similarly you have to make drawing for other two raw materials. Add as much operations as you can, and make two drawings for it as well. So in this way you will have four drawings. So you have to write first two drawings program with G00 and G01 code. And you have to make other two drawings program with cycles only. You only have to use cycles in. It means you will get two programs with G00 and G01 codes and two programs with cycles only. And after writing these programs, you have to develop the jobs on your CNC machine before that, take permission for it, and after that you have to develop these for jobs on CNC machine. So company will give you permission for it as you have been working from long time in that company.

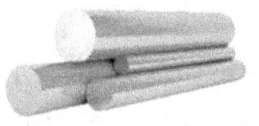

2
- Make Drawings
- Write Program with G00 and G01 codes

2
- Make Drawings
- Write Program with CYCLES

DEVELOP THESE 4 JOBS ON CNC

So you have to learn to develop new job on CNC machine means you have to do setting means firstly do job setting, then select tools, then feed it into rate and then job setting and finally write program on CNC and run. It means you have to do total development and if you are able to do development for these four jobs, then you are s CNC programmer now and you will need only practice in it. So at this stage you have all knowledge that is required to become a programmer. So you are a programmer now.

Now you have to bring perfection in it and need to improve speed of development. So from now you have to start making programs in your company. Firstly, start with small jobs and at this stage as well, you do not have to think about payment. You have to work for next six months in programing, develop as much jobs as you can on CNC machine so you will required first one year to become a programmer and next six months you will require to do practice of programing on CNC machine. After that you have to switch to another company as you have to think about payment now. So find a job in another company as a C and C programmer.

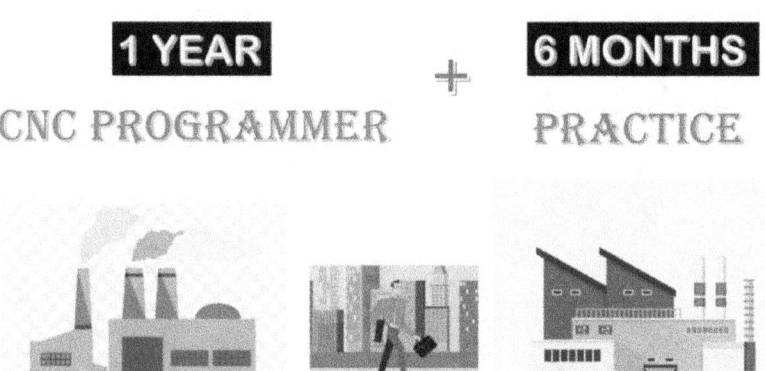

And here also try to get a job in small scale company as you will not be a perfect programmer now, you will still require practice in programing and for that try to find a job in a small company and in that company you have to see which grid is used there and take all the knowledge from that company. And in that company you have to work for next one and half year, make programs for new jobs, develop it and try to bring perfection as well at speed in your work. So you will required one year to become a programmer. Then six months you will require for practice in programing, and then next one and a half a year you work in another company as a C and C programmer means total three years. So now you have total three years of experience. So now again you have to increase your salary. And for that you have to switch to a large scale company as a C and C programmer as you have knowledge as well as experience.

www.ingramcontent.com/pod-product-compliance
Lightning Source LLC
Chambersburg PA
CBHW071020240526
45469CB00006BD/2000